Microbiology Research Advances Series

D1486162

RECENT TRENDS IN BIOTECHNOLOGY AND MICROBIOLOGY

MICROBIOLOGY RESEARCH ADVANCES SERIES

Water Microbiology: Types, Analyses and Disease-Causing Microorganisms
Andriy Lutsenko and Vasyl Palahniuk (Editors)
2010. ISBN: 978-1-60741-273-1

Recent Trends in Biotechnology and Microbiology
Rajarshi Kumar Gaur, Pradeep Sharma, Raghvendra Pratap, Kanti Prakash Sharma,
Manshi Sharma, and Rajiv Dwivedi
2010. ISBN: 978-1-60876-666-6

Microbiology Research Advances Series

RECENT TRENDS IN BIOTECHNOLOGY AND MICROBIOLOGY

RAJARSHI KUMAR GAUR, PRADEEP SHARMA,
RAGHVENDRA PRATAP, KANTI PRAKASH SHARMA,
MANSHI SHARMA AND RAJIV DWIVEDI
EDITORS

Nova Science Publishers, Inc.
New York

For permission to use material from this book please contact us:
Telephone 631-231-7269; Fax 631-231-8175
Web Site: http://www.novapublishers.com

NOTICE TO THE READER

LIBRARY OF CONGRESS CATALOGING-IN-PUBLICATION DATA

Available upon request

ISBN: 978-1-60876-666-6

Published by Nova Science Publishers, Inc. ✢ *New York*

CONTENTS

PREFACE

Biotechnology is one of the major technologies of the twenty-first century. Its wide-ranging, multi-disciplinary activities include recombinant DNA techniques, cloning and the application of microbiology for the production of goods from bread to antibiotics. In this edited book, we provide a complete overview of biotechnology. The fundamental principles of all biotechnology are explained and a full range of examples are discussed to show how these principles are applied; from starting substrate to final product. With the use of comparative studies, this book also discusses the legal, agribusiness and public policy issues that connect intellectual property protection with advancements in agricultural biotechnology.

The purpose of this book is to provide researcher with modern, balanced and thorough knowledge of biotechnology and microbiology. With such a broad array of potential topics and techniques available, it is difficult to select those that researcher should experience and master. The challenge in editing is to pick up the concepts that are most important for our understanding. The problem in editing it is to see where these concepts have changed or new ones have emerged. A large part of the text has been radically written to take account of recent advances. Some chapters have been dismantled and the material from them integrated into other chapters, emphasizing that these topics are not mere side issues for specialists, but belong to the core of biotechnology and microbiology.

We offer this work in good faith and in the hope that it will assist those individuals who seek to gain an understanding of the terminology as it is currently used. However, the reader should be aware that the field of biotechnology is rapidly expanding and evolving, and that new terms are entering the mainstream nomenclature at a rapid pace. In fact, the exact meaning of some of these terms is still under thought, while the meaning of others will undoubtedly be expanded or narrowed as the technology develops. Although we have endeavored to be as accurate as possible, this work is meant to provide a general introduction rather than to be absolute and legally definitive.

We owe a debt to our faculty members and Prof. Shakti Baijal, Dean, FASC, MITS who has generously given advice, comments and correction and played a crucial role in making the text clear, coherent and accessible to researchers.

Editors

Chapter 1 - In spite of the enormous improvements in agriculture, animal husbandry, fisheries and conditions of living in general, a major portion of the world's population still suffer from so called hidden hunger due to food shortage and malnutrition. The increasing population and urbanization has aggravated this problem. Hence, one of the most important issues for all the Governments of the world is to know, how to organize and maintain safe and secure food supply to all its growing number of households for a reasonable price? According to WHO, nearly 30% of the world population suffers from some kind of malnutrition and most of them are from the developing countries of Asia, Latin America and Africa. Twelve million children die due to malnutrition, diarrhoea and other health problems every year. More than half the world's disease burden can be attributed to hidden hunger; unbalanced protein/energy intake or vitamin and mineral deficiencies and developing nations are quickly joining the ranks of countries dealing with severe health issues. Population growth, urbanisation and low purchasing power together with unemployment and bad housing sanitary conditions add to the seriousness of the situation.

Chapter 2 - Microbes are important agents regulating insect populations naturally. The control of insect pests by microbes is distinctive in that naturally occurring host-pathogen relations are manipulated to the benefit of man, for protecting crops or for controlling insect that are vectors of disease. Microbial control agents are presently sorted as alternatives to broad-spectrum chemical insecticides taking into account the many advantages they provide, absence of pesticide residues in food, increased biodiversity in managed ecosystems. Success in identifying microbial agents as components of pest management of horticultural crops is dependent on over coming many problems necessitating extensive research. The paper in addition to describing in detail of the microbial agents utilized as a component of insect pest management strategy in horticultural crops also gives an insight on the future researchable issues to strengthen mycoinsecticides as a component of IPM of insect pests of horticulture and allied areas.

Chapter 3 - Plant viruses have a dramatic negative impact on agricultural crop production including vegetable crops throughout the world & consequently, plant pathologists & agronomists have devoted considerable effort toward controlling virus diseases during this century. Prior to advent of genetic engineering, traditional plant breeding methodology was successfully applied to develop resistance to viruses & conventional methods for control of plant virus disease like control of vector population using insecticides, use of virus-free propagating material, appropriate cultural practices, cross protection & use of resistant cultivars. However, each of the above methods has its own drawback. In recent years, the advancements in plant molecular virology have enhanced our underst&ing of viral genome organizations & gene functions. Moreover, genetic engineering of plants for virus resistance has recently provided promising additional strategies for control of vegetable viral disease. At present, the most promising of these has been the expression of coat-protein coding sequences in plants transformed with a coat protein gene. Other potential methods include the expression of anti-sense viral transcripts in transgenic plants, the expression of viral satellite RNAs, RNAs with endo-ribonuclease activity, antiviral antibody genes, & Ribosome inactivating genes in plants.

Chapter 4 - Microarrays have emerged as a powerful gene expression analysis technique that deals with altered phenotypes at gene expression level. Its ability to correlate various molecular events with gene expression diversify its applications to the fields of pharmaceutical industry, clinical laboratories, in drug development, toxicity assessment, and other applications based on the comparison of expression patterns of genes specific to a particular tissue or genes involved in a metabolic pathway.

Chapter 5 - Chilli pepper (*Capsicum annuum*), one of the most important spice and vegetable crops of the world, is grown in an area of 880,000 hectares in India with a production of 1.2 million tonnes (Anonymous, 2005). India is the world's largest producer, consumer and exporter of chillies to a number of countries viz. Sri Lanka, Bangladesh, South Korea, USA, Germany, Japan, UK and France. In India, Andhra Pradesh, Orissa, Maharashtra, West Bengal, Karnataka, Rajasthan, Tamil Nadu and Uttar Pradesh are the major chilli growing states. The crop is grown throughout the year and in some places overlapping crops are also taken up. Such a cropping system also provides ideal conditions for the perpetuation and spread of viruses and their vectors. It is, therefore, natural that plant viruses have emerged as a major constraint in improving the production of chilli.

Chapter 6 - Stone fruits includes almond, apricot, cherry, nectarine, peach and plum and these are grown on commercial scale in the North and North-Eastern States of India. Losses in terms of quality and quantity are reported from all over the world due the viral infections in these fruits. Important viruses infecting these fruits are: *Apple mosaic virus* (ApMV), *Prunus necrotic ringspot virus* (PNRSV), *Prune dwarf virus* (PDV), *American plum line pattern virus* (APLPV), *Plum pox virus* (PPV), *Apple chlorotic leaf spot virus* (ACLSV), *Apricot latent virus* (ApLV) and Plum bark necrosis stem pitting associated virus (PBNSPaV). In this chapter major virus infecting stone fruits has been described i.e. *Prunus necrotic ring spot virus*. Latest biotechnological approaches used in the management practices like virus free production of plants, CP mediated resistance and RNAi used to manage viral diseases are also described.

Chapter 7 - The possibility to transfer genes across almost all taxonomic borders by molecular techniques has expanded the potential resources available to plant breeders enormously. It is becoming generally accepted that a multidisciplinary approach to plant biology will lead to the disappearance of borders between disciplines. In the same vein, the differences between transgenic and non-transgenic crops should become irrelevant when the focus of plant breeding is on achieving maximal production in a sustainable way to feed the growing human population.

Dubbed 'the Green Phoenix', the transgenic plant technology offers both challenges and opportunities for growth and development of mankind. This technology should be used to complement the traditional methods for enhancing productivity and quality, rather than to replace the conventional methods. To adopt this technology, GM crops and their products, awareness has to be created among the farming and consumer communities regarding their benefits and effects on human life, by the scientific communities and national leaders. This presentation also includes a few recent advancements in transgenic plant technology and their potential role in enhancing the quality of human life.

Chapter 8 - Phytoplasmas cause diseases in several weeds which may act as alternative natural hosts facilitating the spread of phytoplasmas to other economically important plants and thereby increasing economic losses. The most peculiar symptoms observed on weeds include extensive chlorosis, proliferation of axillary shoots, witches'-broom, yellowing and little leaves. So far more than 43 weed species were reported having phytoplasma infections from all over the world. Nucleotide sequence studies have shown that weed-infecting phytoplasmas mainly belongs to 16SrI, 16SrIV, 16SrVIII, 16SrXI and 16SrXIV groups. Among them, 16SrI and 16SrXIV phytoplasmas have a more wide occurrence in nature all over the world. Even though the weeds identified as phytoplasma hosts often grow abundantly around field crops, the possibilities of transmission of phytoplasmas related to important agricultural, economical and horticultural crops from weed to the mentioned crops and vice-versa can not be ignored. This could be because phytoplasmas are able to survive in many potential economical crops or because an insect vector is capable of transmitting phytoplasmas from other weeds to crops which are already known as phytoplasma hosts. In either case, the chance of transmission in the future seems high, given the large phytoplasma reservoir already revealed, the propensity of new phytoplasma strains to evolve. In this chapter, detailed and up-to-date information on occurrence, symtomatology, molecular characterization, transmission, taxonomy, genetic diversity and management approaches on weed phytoplasmas has been discussed. Knowledge of the diversity of phytoplasmas will be expanded by recent studies and the availability of molecular tools for pathogen identification.

Chapter 9 - The ability of suppress transiently mRNA accumulation of specific genes in a high-throughput is a powerful tool in a genomics-scale approach to assign biological function to uncharacterized genes. Virus induced gene silencing, or VIGS, is a method o transiently interrupt gene function through RNA interference. The exact mechanism by which VIGS operates is still unclear. It is known, however, that this approach harnesses the plants natural ability to suppress the accumulation of foreign RNAs by an RNA-mediated defense mechanism against plant virus. The systemic signal by which this mechanism is induced is unknown, but it is thought to involve an RNA component. Recently it was shown that inoculation of transcript from cloned viruses capable of expressing host sequences in plants led to silencing of the homologous host gene. The infected plants displayed a phenotype representative of the loss of function of the host gene, and not of virus infection.

Chapter 10 - various biological phenomena in plants. A molecular marker is a particular segment of deoxyribonucleic acid (DNA) that is representative of the differences at the genome level. As genomes of all plants cannot be sequenced, therefore molecular markers and their correlation to phenotypes provide requisite landmarks for elucidation of genetic variation. Genetic or DNA based marker techniques such as restriction fragment length polymorphism (RFLP), random amplified polymorphic DNA (RAPD), simple sequence repeats (SSR) and amplified fragment length polymorphism (AFLP) are routinely being used in evolutionary, phylogenic and genetic studies of plants. AFLP technique combines the power of RFLP with the flexibility of PCR-based technology by ligating primer recognition sequences (adaptors) to the restricted DNA and selective PCR amplification of restriction fragments using a limited set of primers. The AFLP technique generates fingerprints of any DNA regardless of its source, and without any prior knowledge of DNA sequence. Most AFLP fragments correspond to unique positions on the genome and hence can be exploited as landmarks in genetic and physical mapping. The technique can be used to distinguish closely related individuals at the sub-species level and can also map genes. Significant progress in crop productivity has been made in India due to painstaking efforts of plant breeders; however future possibilities of crop improvement include development of new and more efficient plant ideotypes with improved quality traits. This objective can be successfully achieved, if conventional plant breeding is supplemented with molecular breeding approaches i.e. including both the transgenic crops and the marker-assisted selection (MAS).

Chapter 11 - Mycorrhiza means when fungi enter into a mutualistic relationship with the plant roots. In this type of relationship, fungi actually become integrated into the physical structure of the roots. This sort of association helps both the partners i.e. fungus and its host plant to be mutually benefited by each other as the fungus absorbs water and nutrients from soil and supplies the same to the plant and in turn derives its nutrition (carbohydrates and photosynthates) from the host plant for its growth and multiplication.

Chapter 12 - India has a variety of soils and climates. Therefore, almost all kinds of fruits and vegetables, *viz.*, temperate, sub-tropical and tropical are being grown in different agroclimatic regions of the country. This natural advantage promises great potential for the country's fruit and vegetable culture. Development of good fruit and vegetable industry on sound scientific basis could be of benefit in many ways. They enrich human diet by supplementing vitamins, minerals and sugars in addition to being an easily digestible food.

The importance of fruits and vegetables for our country, which is facing acute shortage of food can neither be denied nor disparaged. It is unfortunate that country suffers a great loss due to considerable damage caused to fruits and vegetables by a number of diseases occurring during post-harvest period (transit and storage). These are highly perishable things and the losses are more considerable than as often realized, because fruits and vegetables increase manifold in unit value while passing from the field at harvest to the consumer. There are more than 250 known parasitic diseases of fruits and vegetables that cause decay and blemishes during transit, marketing and storage. The damage and losses incurred vary with the crop, growing conditions in the field, handling during post-harvest and transit, and storage conditions. In a developed country like U.S.A., where advanced post-harvest technology is applied, annual loss of fruits and vegetables is approximately to the tune of 200 million dollars. According to an old estimate New York city alone suffers losses of 700 car loads of fruits and vegetables every year. In India exact data on losses are not available, however, the data collected from some past studies put the average loss of fruits and vegetables at 20-30 percent (Bose *et al.*, 1993).

Chapter 13 - HIV infections are always feared, as its infection can only be controlled and can not be cured. It is almost more than 25 years have gone by with extensive researches on HIV; we failed to develop either any vaccine or a suitable therapy to cure HIV.

Undoubtedly, developments of better anti-retroviral drugs as well as improvement in treatment regimens have significantly prolonged survival among HIV patients. These developments have surely helped HIV seropositives to lead a better and longer life; but, it has certain drawbacks, too. The major draw back is increase in incidences of HIV associated dementia (HAD) and other neurological disorders. These can be commonly defined as NeuroAIDS. The worrying part of this whole saga is that NeuroAIDS affects its patients in their prime of life *i.e.*, ~30-40 years of age.

In India, numbers of HIV seropositives are very high (~2.5 millions). In the recent past, due to sincere efforts by the Indian Government, NACO, UNAIDS and various NGOs, now these patients have better access to improved anti-retroviral treatments. These efforts will help to prolong life-expectancy among HIV seropositives, so that they can lead a quality life. Besides this, high pathogenicity of HIV strain prevalent in India may also be a reason for concern on its impact on HAD.

When we put all these facts together, it definitely suggests that soon India will witness an enormous increase in number of NeuroAIDS patients. It is a high time to initiate appropriate measures to control this upcoming health problem in India.

Chapter 14 - *Gloriosa superba* L., a member of the Liliaceae family is locally called as "Kalihari". Plant produces alkaloids, mainly colchicine and colchicoside. The derivatives of colchicines have shown promising anti-cancer and anti-inflammatory properties. The normal regeneration and propagation of *Gloriosa superba* is through corm are slow and insufficient for conservation. Poor seed germination, susceptibility towards many pests, collection of seeds by experts for pharmaceutical industries and reduction in forest area are important factors responsible for diminishing population. Seeds of *G. superba* germinated on MS medium + 0.1 mg/l BAP. The activated embryo exhibited callus induction within 23-25 days from inoculation on MS medium + 2,4-D (2.0 mg/l). MS medium supplemented with 1.0 mg/l 2,4-D + 0.5 mg/l Kinetin + 0.5% Maltose + additives was found to be the best medium for multiplication of cell cultures. On MS + 1.0 mg/l BAP + 0.5 mg/l Kinetin callus turned embryoidogenic. On MS medium + 2.0 mg/l BAP + 0.5 mg/l Kinetin the callus showed rhizogenesis. The methodology developed and defined is reproducible and can be utilized for production of active constituent (colchicine) using plant biotechnology tools.

Chapter 15 - The manipulation of living beings has undergone a real revolution with the arrival of genetic engineering, which allows part of the gene pool to be isolated and manipulated as desired. Thus, genetic engineering has led to the appearance of "genetically modified organisms" or GMOs. A Genetically modified organism (GMO) or genetically engineered organism (GEO) is an organism whose genetic material has been altered using genetic engineering techniques. These techniques are generally known as recombinant DNA technology. With the recombinant DNA technology, DNA molecules from different sources are combined in vitro into one molecule to create a new gene. This DNA is then transferred into an organism and causes the expression of modified or novel traits. Contemporary genetic modification was developed in the 1970s and essentially transfers genetic material from one organism to another.

Chapter 16 - Today we are living in the age of Information Technology and spreading our wings to every aspect of the society. We exchange our information, as it is the human tendency to share their innovative thoughts with their near and dear ones but that proves fatal, in most of the cases, in respect of that original creation or property i.e. Intellectual Property (IP). IP laws are extremely important for the scientific development, growth and prosperity of any country. The international community realized this around the early 1990's and fallout of this realization came out in the form of Trade Related Aspects of Intellectual Property Rights (TRIPs) agreement in 1994 and its enforcement on 1st January 1995. A patent is an exclusive right granted to inventor or creator of a useful or improved article or a new process of making an article for a specified period of time.

India is a member of the World Trade Organization (WTO) and a signatory to the WTO agreement on TRIPs. Pursuant to the TRIPs agreement, India has amended its patent legislation which is governed by the Patent Act, 1970 on three occasions. The case of Diamond V. Chakrabarty in 1980 led to general trend of patenting inventions on living matter. In India, the position was made clear after the 2002 amendment to the Indian Patents Act. The amended act stated that life forms can be patented provided they satisfy the other requirements. The improvements in the Indian patent regime have resulted in a significant up thrust in the promulgation and enforcement of patents in India. Now, India can boast of one of the best patent law regimes in the world. However, a lot remains to be achieved still. Improving IPR protection will be an important element to increase the attraction of private investors in India.

Chapter 17 - The biotechnology industry is one of the most research and development intensive and capital-focussed industries in the world. As such, its success relies on a robust intellectual property system as well as a strong set of competition laws. The management of biological resources has been an increasingly contentious subject at the national and international levels. This is linked in large part to the progressive recognition of new economic opportunities arising from the use of biodiversity. As a result, international legal frameworks for the management of biological resource have had to increasingly take into account not only the needs of biodiversity conservation but also concerns about its potential for economic use and its contribution to the process of economic development. This has important repercussions from a legal perspective because the new products developed by the biotechnology industry can often easily be copied once they have been put on the market. As a result, the biotechnology industry has strongly argued for the introduction of Intellectual Property Rights (IPR) over genetically modified organisms. This paper seeks to analyze the impact of the international legal framework for the promotion of IPR on India's legal regime concerning the control over biological resources and inventions derived from biological resources. It focuses in particular on three acts and legislative amendments adopted in recent years and their organic relationship within the overall domestic legal framework. The paper analyses these enactments in the context of the move towards the control of biological resources and derived products through IPR. This has ramifications not only for control over biological resources and derived products but also more generally on the management of agriculture in India and other developing countries and the realization of food security and the human right to food at the individual level.

"These varieties have 50 percent higher yields, mature 30 to 50 days earlier, are substantially richer in protein; are far more disease and drought tolerant, resist insect pests and can even out-compete weeds. And they will be especially useful because they can be grown without fertilizer or herbicides, which many poor farmers can't afford anyway. This initiative shows the enormous potential of biotech to improve food security in Africa, Asia and Latin America."

In: Recent Trends in Biotechnology and Microbiology   ISBN: 978-1-60876-666-6
Editors: Rajarshi Kumar Gaur et al., pp. 1   2010 Nova Science Publishers, Inc.

FOREWORD

My experience of teaching and research has stimulated a deep interest, insight and approach to the practical and spiritual affairs of our day-to day life. In such a pious quest; be it for science, literature, politics or religion, one desires a fundamental knowledge to explore a rational relationship of the subject with one's welfare. This book meets that zeal and proves the basic science to act as a tool in our life. As a witness of this book's birth, I have had no fret for its significance and worthiness among students and teachers, and of course for all.

The department of science, FASC-MITS University has really congregated intellectuals to reveal the forthcoming human ailments caused by malnutrition and drought, as well as annihilation of potable water. This book is a threat-in-advance before it is too late, for the sustainable survival of society. Today we have distanced ourselves from nature and arrived at a bank of extinction; fields have dried up, water disappeared, cuckoos hunted down, grass land shrinked, inorganic food, moral values closed in books only, etc. draw a panoramic portrayal of modern time. The whole land has turned into 'wasteland', no avenue, only rocky path; the pond of faith in nature has also dried up. All these images paint a horrible time for us. It is quite timely to awaken the people to form-

"Same lush green Park,
In the sky same Skylark".

I do not believe in romanticism, but in reality that is very sour today. A renaissance spirit is urged to make this reality sweet. The present book in your hand promises a strong guarantee for the overall safe survival. It has, as intimated to me, touched the fine grounds needed for all of us on this planet earth.

To return to the main purpose of this book, I have a very high opinion of its utility. Literally observed, it may prove a boon to humanity and the mother earth.

Ashok Singh Rao
Head Department of Arts
FASC, MITS

In: Recent Trends in Biotechnology and Microbiology
Editors: Rajarshi Kumar Gaur et al., pp. 3-7

ISBN: 978-1-60876-666-6
2010 Nova Science Publishers, Inc.

Chapter 1

FERMENTED FOODS IN HUMAN WELFARE

*J.B. Prajapati**

Coordinator, Swedish South Asian Network on Fermented Foods; Professor and Head,
Department of Dairy Microbiology, SMC College of Dairy Science, Anand Agricultural
University, Anand 388 110, Gujarat State, India.

ABSTRACT

In spite of the enormous improvements in agriculture, animal husbandry, fisheries
and conditions of living in general, a major portion of the world's population still suffer
from so called hidden hunger due to food shortage and malnutrition. The increasing
population and urbanization has aggravated this problem. Hence, one of the most
important issues for all the Governments of the world is to know, how to organize and
maintain safe and secure food supply to all its growing number of households for a
reasonable price? According to WHO, nearly 30% of the world population suffers from
some kind of malnutrition and most of them are from the developing countries of Asia,
Latin America and Africa. Twelve million children die due to malnutrition, diarrhoea and
other health problems every year. More than half the world's disease burden can be
attributed to hidden hunger; unbalanced protein/energy intake or vitamin and mineral
deficiencies and developing nations are quickly joining the ranks of countries dealing
with severe health issues. Population growth, urbanisation and low purchasing power
together with unemployment and bad housing sanitary conditions add to the seriousness
of the situation.

INTRODUCTION

In spite of the fact that the food productions of the world have been doubled during the
last thirty years, house hold food security for a considerable part of the world population has
not been solved satisfactorily. Improper post-harvest handling and lack of infra structure for
processing and distribution of foods are also another contributing factors.

*E-mail: prajapatijashbhai@yahoo.com

Research and higher education in food science and biotechnology could make an important contribution in solving this problem, by finding out the ways for efficient utilization of existing resources, developing low cost effective technologies for making the foods more nutritional and palatable and preventing the loss due to biological, microbiological, chemical, biochemical, mechanical, physical and physiological factors.

One of the possible approaches in finding solutions to these problems is exploiting the full potential of fermentation technology. Fermentation is one of the oldest methods of food preservation known to mankind which is applied globally (Prajapati and Nair, 2003). Several kinds of indigenous fermented foods have come-up from various parts of the world since centuries and are strongly linked to culture and tradition, especially in rural households and village communities (Farnworth, 2004). It is estimated that fermented foods contribute to about one-third of the diet worldwide (FAO, 2004; Nout and Motarjemi, 1997).

The reasons, why we propose fermentation technology as a means of tackling problem of malnutrition and ill-health can be summarized as under.

1) It has strong cultural foundation as various kinds of fermented foods have been reported from all most every part of world.
2) It gives excellent exposition of biodiversity as it can be applied to a wide variety of raw materials to produce a variety of different finished food products. Cereals, pulses, root crops, vegetables, fruits, milk, meat and fish are preserved by one or other method of fermentation.
3) The method is very simple and environment friendly
4) It is less energy consuming and produce less waste
5) It require less and simple equipment and is gender friendly
6) Easy to manage under house hold conditions of low income communities as well as in industrial scale for urbanised welfare societies
7) The process improves the sensory properties of the foods like taste and smell as well as body and texture properties.
8) It improves the safety, as the microflora bringing out fermentation can check the growth of pathogens.
9) The fermentation technology leads to improvement of nutritional properties of food by starch degradation, protein digestibility, production of vitamins, liberation of enzymes and degradation of anti-nutrients like phytic acid.
10) The use of fermented milks in control of intestinal disorders is well known and several other therapeutic benefits like; anticarcinogenic properties, hypocholesterolemic action, immunostimulating activities, etc are showing encouraging results.

Apart from all these, it offers immense opportunity for production of products, which can be classified as functional foods, organic foods, natural foods, health foods, convenience foods, ethnic foods, neutraceuticals, foods for clinical nutrition, probiotics, prebiotics and synbiotics.

WHAT WE NEED TO DO?

The potential of fermented foods can be exploited for the benefit of large masses in the third world who are suffering from mal-nutrition and poor health. However, this requires sincere efforts from all the stakeholders. The first requirement is to educate the common man about the health benefits of fermented foods and promote traditional fermented foods to the fullest extent. Media as well as social and health workers can play an important role in this campaign.

To manufacture fermented foods with improved technology and selected probiotic bacteria requires systematic research and development. This needs sincere efforts by academic institutions and the food industry. Some of the R & D needs in this area are;

1) Selection of raw materials for fermentations
2) Understanding the course of fermentation
3) Types, numbers and changes in microflora
4) Optimization of Biochemical changes
5) Assessment of Nutritional improvements
6) Clinical trials to evaluate therapeutic properties
7) Technology for preservation, packaging and distribution

R & D is considered to be a costly item in the economy of any food company. However research and development could be effective and less expensive when it is carried out as collaborative network projects. Development of new foods and testing its special health benefits need input from many specialized areas of technology and science including agriculture, food processing, microbiology, biochemistry, nutrition, medicine, sociology, extension, etc. The SASNET-Fermented Foods has taken the initiative in this direction to build a network of all stakeholders who can promote R & D in fermented foods and propagate the use of fermented foods on wider scale (Nair et al, 2006).

Further, for reaching up to the last person in the society, there is need to evolve a special system or alternatively, some of the existing systems can be utilized. For example, the mid day meal program for school going children is running successfully in many States. This programme can conveniently include suitably packed products which are designed to take care of nutritional enrichment and health promotion.

OUR RESEARCH INITIATIVE

Looking to the multifarious benefits of fermentation technology, initiative has been taken at our institute to isolate and characterize important lactic acid bacteria (LAB) for fermentation of milk and employ them in manufacture of fermented milk based products. Few examples of the technologies for manufacture of such foods are given hereunder.

Acidophilus Powders

The liquid fermented milk contains vigorously active and high number of lactobacilli in it, but due to its low shelf-life and marketing inconvenience, it has not become commercially popular. Dried products have inherent advantages of long term shelf-life, low volume, easy transportation and convenience in handling. However, the main task is to select suitable carrier cum protective media for probiotic cultures and apply suitable technology for its drying, to have higher number of viable cells. Keeping these challenges in mind, attempts were made at our laboratory to manufacture low cost dried acidophilus products.

In one process, acidophilus milk made with process adapted human strain of *Lactobacillus acidophilus,* was blended with banana paste, tomato juice and sugar and finally spray dried. The powder product was organoleptically acceptable and had higher nutritive value in terms of vitamin content and protein quality. When it was fed to human volunteers, the lactobacilli could implant in the intestinal tract suppressing the population of coliforms. The product showed as high as 14% survival of lactobacillus culture (8×10^7 cfu/g), which reduced to about one log cycle, during 2 months of ambient storage (Prajapati *et al*, 1986).

Another product was standardized from a mix containing malted wheat flour, acidophilus milk, sugar, cocoa and salt. This spray-dried product contained about 10 million live *Lb. acidophilus* cells per gram and was nutritionally superior and organoleptically appealing. The product had 17% protein, 2% fat, 75% carbohydrates and 3% ash (Shah *et al*, 1987).

Milk-Cereal Based Powder

Development of milk-cereal blends has dual advantages. Firstly, the cost of the product is reduced by replacing a part of costly milk solids with cereal solids and secondly, the nutritive value is enhanced due to the complimentary action of milk and cereal nutrients. The health benefits of such blends can be further enjoyed when they are added with suitable probiotic culture. We attempted to develop blends of milk and cereals by trying several combinations of milk, butter milk, whey, rice and wheat and appropriate thermal and enzymatic treatments were standardized to make the products suitable for spray drying and improving their organoleptic and textural properties. Selected milk-rice blend was converted into a probiotic food by incorporating freeze-dried cells of *Lactobacillus acidophilus* V3 in it. The product had on an average 10.9% fat, 22.3% protein, 3.2% ash, 61.7% carbohydrate and 2.1×10^9 cfu/g of live lactobacilli (Prajapati, 2003).

Milk-Fruit-Vegetable Based Synbiotics

Milk fermented by probiotic lactobacilli and added with prebiotic ingredients like inulin, have been blended with shreds of banana, sapota, cucumber and carrots. Such products are readily acceptable in Indian diet as they are known traditionally as *raita* and used in daily diet. These products can become a good delivery vehicle for probiotics.

EPILOGUE

Fermented foods offer the solution to the problem of malnutrition and ill-heath in the developing world. It has become a matter of great interest to scientists, medical practitioners, food companies, and marketing agencies as the demand for such foods is enormous and it is growing at a rapid rate. There is lot of scope to enhance use traditional fermented foods and manufacture novel foods with health bacteria on an industrial scale. By any means, propagation of fermented foods on wider scale can help in getting closer to the UN millennium goal of reducing under-nutrition by half by 2015.

REFERENCES

[1]   FAO. (2004). Biotechnology applications in food processing: Can developing countries benefit? *Electronic forum on Biotechnology in Food and Agriculture*: Conference 11.

[2]   Farnworth, E.R. (2004). The beneficial health effects of fermented foods- Potential probiotics around the world. *J. Neutraceuticals, Functional and Medical Foods*, 4(3/4):93-117.

[3]   Nair, B.M., Prajapati, J.B. & Varshneya, M.C. (2006). SASNET-Fermented foods: An initiative for meeting the challenges of poverty and hunger. *Indian Food Industry*, 25(5):13-15.

[4]   Nout, M.J.R. & Motarjemi, Y. (1997). Assessment of fermentation as a household technology for improving food safety: a joint FAO/WHO workshop. *Food Control*, 8:221-226.

[5]   Prajapati, J.B. (2003). Technologies for manufacture of milk based fermented foods" *Invited talk at National Workshop on "Identification of technologies and equipment for meat and milk products"*, sponsored by Department of Science & Technology, Govt of India at Indian Veterinary Research Institute, Izatnagar, Bareilly, September 5-6, 2003, pp. 136-137.

[6]   Prajapati, J.B. (2008). Fermented foods as a tool to combat malnutrition. *In "Food and Water Security"* edited by U. Aswathanarayana, Taylor & Francis, London, pp. 159-168.

[7]   Prajapati, J.B. & Nair, B.M. (2003). The history of fermented foods. *In "Fermented Functional foods"* edited by Edward R. Farnworth, CRC Press, Boca Raton, New York, London, Washington DC, pp. 1-25.

[8]   Prajapati, J.B., Shah, R.K. & Dave, J.M. (1986). Nutritional and Therapeutic benefits of a blended-spray dried acidophilus preparation. *Cult. Dairy Prod. J.* 21 (2):16-21.

[9]   Shah, R.K., Prajapati, J.B. & Dave, J.M. (1987). Packaging materials to store a spray dried acidophilus malt preparation. *Indian J. Dairy Sci.* 40(2):287-291.

In: Recent Trends in Biotechnology and Microbiology
Editors: Rajarshi Kumar Gaur et al., pp. 9-23

ISBN: 978-1-60876-666-6
2010 Nova Science Publishers, Inc.

Chapter 2

MICROBES AS A TOOL IN INSECT PEST MANAGEMENT OF HORTICULTURAL CROPS IN INDIA: PROGRESS AND PROSPECTS

P.N. Ganga Visalakshy

Division of Entomology and Nematology, Indian Institute of Horticultural Research, Bangalore.

ABSTRACT

Microbes are important agents regulating insect populations naturally. The control of insect pests by microbes is distinctive in that naturally occurring host-pathogen relations are manipulated to the benefit of man, for protecting crops or for controlling insect that are vectors of disease. Microbial control agents are presently sorted as alternatives to broad-spectrum chemical insecticides taking into account the many advantages they provide, absence of pesticide residues in food, increased biodiversity in managed ecosystems. Success in identifying microbial agents as components of pest management of horticultural crops is dependent on over coming many problems necessitating extensive research. The paper in addition to describing in detail of the microbial agents utilized as a component of insect pest management strategy in horticultural crops also gives an insight on the future researchable issues to strengthen mycoinsecticides as a component of IPM of insect pests of horticulture and allied areas.

INTRODUCTION

India is the second largest producer of horticulture of the world. Horticulture being labour intensive generates employment and provides nutrient for the people. Hence, horticulture plays a vital role in the prosperity of a nation and is directly linked with the health and happiness of the people. One of the major factors that hinder horticulture production is the damage caused by insect pests.

Crop protection still, is based largely on the use of chemical pesticides. Use of chemical pesticides has led to varied problems such as pest resistance, resurgence of pests, presence of chemical pesticide residues in end products, elimination of natural enemies of pests, disruption of ecological balance, causing environmental degradation /pollution and health hazards to all kinds of animals including human beings. This has led to the demand for management of insects pests by alternate non- chemical methods such as biological control, botanicals etc that are effective and safe to all other forms of living things in the universe.

Today biological control has expanded from the use of entomophagous insects to the use of a whole range of organisms to control insect pests. Microbial control agents inimical to pests represent an ideal form of pest control, particularly for use in both short and long term pest suppression. They are ecologically less disruptive and highly safe to humans and animals.

The procedures used to achieve microbial control include the intentional creation of epizootics or the utilization of naturally occurring epizootics [Harper, 1987].Among the various microbes developed and tested, virus, bacteria, protozoan, fungi, and nematodes are considered promising. The success of microbial control agents is based on the many different interactions between the entomo-pathogens and host insects. The development and spread of the disease within an insect population is dependent on the interaction of the pathogens, the host and the environment that are variable and complicated. This may result in additive, synergistic or antagonistic leading to an over all effect on the insect which may vary from lethal to sub lethal effect (Burghes, 1981).

Today microbial agents are considered important sources of raw materials for biotechnology and especially genetic engineering. The advances in biotechnology have proved to be a potential tool capable of solving some of the complicated problems which other wise could not be solved. Its possible use as an additional IPM component strengthens the exiting crop protection methods (Datta, 1992). The present paper details on the status of microbes in insect pest management of horticultural crops in India and the future research needs in the area of microbial control that could form an important component in integrated insect pest management in Horticulture.

MICROBES IN HORTICULTURE INSECT PEST MANAGEMENT

Insect Viruses

Work on viruses in India was initiated as early as 1968 with the report of nuclear polyhedrosis virus (NPV) from *Helicoverpa armigera* (Patel *et.al*.1968), a pest of national importance and *Spodoptera litura* (Ramakrishnan and Tewari, 1969, Dhandapani *et.al*.1992) a pest of many horticultural crops. Since then studies on insect viruses have progressed rapidly and several viruses were reported to occur in insect pests most of them from order Lepidoptera. These comprise of nuclear polyhedrosis virus, (NPV), granulosis virus (GV) and cytoplasmic virus (CPV).

NPV

Of the different type of viruses, NPV has the greatest potential since they are virulent, faster in action than GV and CPV. The NPV of *H.armigera* has been studied extensively to develop it as viral pesticide. *Ha* NPV, a single embedded virion type has a high virulence against *H.armigera* infesting horticultural crops such as tomato, field bean, pigeon pea, grape (Narayanan and Gopalakrishnana, 1988]. *Spodoptera litura* is also a major pest of crops such as cabbage, Chilli. and grapes. An NPV isolated from this insect is reported to effectively control the pest in crops such as chilli (Dhandapani and Jayaraj, 1989) cruciferous crops (Sharma and Chaudhary, 1994).

In addition to these crops NPV are reported from brinjal shoot and fruit borer (Tripathi and Singh, 1991), pumpkin caterpillar, *Diaphania indica*. (Narayanan and Veena, 2002) Cabbage butterfly, *Pieris brassicae* (Battu, 1995) *Crocidolomia binotalis* (Battu, 1989), Pomegranate Hairy caterpillar, *Trabala vishnou* Lef (Mani *et al.*, 2000). Ber Hairy caterpillar, *Thiacidas postica* Walk (Mani *et al.*, 2001) and Granolosis virus from DBM on cabbage (Rabindra *et.al.*, 1998). However, detailed research on their field efficacy for developing as a commercial formulation is to be carried out.

Bacterial Pathogens *Bacillus Thuringiensis*

Bacterial insecticides are manipulated maximum for management of insect pests. Of the various species *Bacillus thuringiensis* Berliner has received considerable attention from insect pathologists, economic entomologists and commercial producers of pesticides. *B.thurigiensis* with its varieties belongs to a group of crystalliferous spore formers (Bt end toxin) found in the sporangium with spore. The fact that *B. thuringiensis* can be easily mass cultured on laboratory scale as well as on industrial scale confer to this pathogen great potentialities in microbial control of insect pests. The wide host range, relative susceptibility of the toxic crystal and the speed and the ease in mass production on wide variety of media have led to its exploration as a basis for a number of microbial insecticides.

In India, the bacterial insecticide has been used in the management of several horticultural pests notably, cabbage butterfly *Pieris brassicae* L. on cruciferous crops (Atwal and Singh, 1969). *S.litura* on cabbage (Datta and Sharma,1997), fruit borers of okra, brinjal, tomato (Sathpathy and Panda,1997 Krishnaiah, *et al.*, 1981), DBM on cabbage, bud borer of cucurbits, fruit borer in ber and pomegranate, citrus leaf miner, *Phyllocnistis citrella* (Staint.) and citrus butterflies *Papilio demoleus* L. and *Papilio polytes* L (Krishnamoorthy and Ganga visalskhy, 2007, Gopalakrishnan and Ganga Visalakshy, 2005).

However, there are several reports indicating the ineffectiveness of Bt preparations against many lepidopteran pests. Hence, it is important that new strains of Bt with increased host spectrum and virulence be developed. Also as of now all BT formulations in the market are imported. Hence there is an urgent need to isolate effective indigenous strains with high virulence, develop cheap mass production methods and stable formulations and patent them. Since insects are reported to develop resistance, this biopesticide should be used judiciously (Gelernter, 1997).

Fungal Pathogens

Fungal diseases in insects are common and wide spread and often decimate populations in spectacular epizootics. All insect order is susceptible to fungal diseases. Nevertheless of the 700 species of entomopathogenic fungi currently known, only a dozen have been exploited up to commercial products. It was in 1950's that the East European countries started investigations particularly with *Beauveria bassiana* as a part of general strategy to control the Colorado potato beetle *Leptinotarsa decimlineata*. This led to the practical approach of utilizing fungal pathogens as part of an integral control strategy of insect pests.

Though all groups of fungi cause infections to insects, the Deutromycetes (Entomotphora, Coelomycetes, and Massospora and genus *Beauveria, Metarhizium, Verticillum, Nomurea, Peacilomyces, Asprgillus, Hirustella, Ashersonia* and Basidomycetes are the most important ones for insect control.

Many entomo-pathogens are reported to cause epizootic against horticultural crops. The fungus *Nomuraea rileyi* (Farlow) Samson on *H.armigera* on tomato (Gopalakrishnan and Narayanan, 1989), *S.litura* and *S.exigua* (Phadke, *et.al.*,1978), *Bipolaris tetramera* (Mckiney) Shoemaker and *Beauveria bassiana* on brinjal fruit borer (Tripathi and Singh 1991, Anonymous,2006), *Paecilomyces farinosus* (Holmskiold) Brown & Smith and *Zoophthora radicans* (Brefeld) on *P. xylostella, Verticilum lecanii* from thrips (Ganga Visalakshy,*et.al.*,2004), *Metarhizium anisoplaie* from hoppers, *Beauveria bassiana* from thrips, *Entomopthora* from hopper, *Fusarium monolifera* from thrips (Anonymous, 2006,2007 and 2008), cause epizootics under favourable conditions. These pathogens were exploited for suppressing the pest under natural conditions and promising results were obtained by using *Nomurea rileyi* against *H.armigera* and *Paecilomycis farinosus* against DBM on cabbage, *M.anisoplaie, Verticillum lecanii* and *Beauveria bassiana* against *Scitothrips dorsalis* on chilli and capsicum and *M. anisopliae* against mango hopper, *Idioscopus nitidulus* (Ganga Visalskhy, *et. al.* 2009, Sitandanum,*et.al.*2007, Anonymous,2007,Ganga Visalakshy, *et. al*, 2007b, Mani *et.al.*2005).

Protozoan Pathogens

Though they are naturally inflicting affecting natural populations, protozoans are given very little importance as applied microbial agents of insect pests as they cause only chronic infection in a narrow range of hosts and not effective as compared to other agents. Some of the protozoan pathogens are the Pebrine disease caused by *Nosema bombycis* in silkworm, *Nosema apis* in honey bees (Lawrence lacey, 1997).

Entomophilic Nematodes (EPN)

Nematodes that are parasitic on insects form effective control agents. These agents are very easy to multiply, effective, sustainable compatible with other plant protection measures etc. EPN are reported to work better in soil insects. Another advantage is that they are exempted from registration.

EPN's gaining importance, as they possess many positive attributes of potential biological control agents like broad-spectrum effectiveness, short life cycle, amenability to mass production, recycling ability, persistence etc..EPN belonging to family *Steinernerna and Heterorhabditis* were evaluated against various horticultural crop pests such as *S.litura, brinjal borer,* DBM, mustard saw fly *Athalia proxima*, cirus butterfly *Papilo demoleus* and were observed to be effective (Hussaini and Singh,1998,Ganga visalakshy *et.al.* In presss)

The success of controlling insect pests by microbes is dependent on many physio-chemical and biological approaches such as timing of application, dosage, economics, insect feeding behavior, sunlight, rainfall, compatibility with other protection practices. Of late, the biotechnological approaches are also gaining importance in microbial control of insect pests. The ecology, epizootology and population dynamics of insect microbes is an interesting area of research that has received relatively little attention.

PHYSIO-CHEMICAL APPROACHES

Cultural Approaches

Timing of Application
Infection of fungal spores and entomophilic nematodes are from exterior and is dependent on the contact of the spores with the insect. However, in the case of virus and bacteria infection and mortality is dependent on the quantum of spores ingested by the insect. Hence, infection and effectiveness varies with the microbes used. In the case of virus and bacteria as indicated since ingestion is important, timing of application is related to the feeding behavior of the insect. Noctuid insect larvae are active at night. Hence spraying of the microbes such as *Ha* NPV, *Sl* NPV and *Bacillus thuriengenesis* spores at evening hours is recommended to cause higher mortality of the target pests (Narayan and gopalakrishnan, 1987, Asokan and Mohan, 1996, Mani.et.al.2005)-.

The success of *H.armigera* control through the use of NPV has been mainly attributed to the timing of treatment and coverage of plant parts. On tomato, three rounds of *Ha* NPV application @ 250 LE/ha (1.5 x 10^{12} POB/ha) along with adjuvant during the evening hours at weekly intervals right from the flower initiation resulted in significant reduction in the borer damage (Narayanan and Gopalakrishnan, 1987; Mohan *et al.*, 1996; Sivaprakasam, 1998). These rounds are to be sprayed at 25, 32 and 42 days after planting which is the critical period of infestation of the fruit borer. The high reduction in larval population following three applications of NPV at weekly intervals is attributable to the time of synchronization with the infestation of the insect especially occurrence of early instar larval stages.

In addition to the time of application, effectiveness is also dependent on the critical time of infestation. For example in *H.armigera* on tomato, bhendi etc. it is recommended to initiate spraying with flower initiation that coincides with the pest infestation (Narayanan and Gopalakrishnan, 1988). In chilli, spraying of entomo-pathogens against thrips is recommended with the initial signs of damage of the terminal leaves (Pers. Observation). In cabbage, spraying against DBM is recommended with 20 days after planting, which coincides critical period of infestation of DBM (Krishnamoorthy, *et.al.* 2005).

Directional Spraying

Effectiveness also depends on the direction of spraying. It is very important that the microbes have to come in contact with the pest in the case of fungus and EPN while in the case of virus they must be in the area preferred for feeding by the larvae. Therefore spraying of microbes must be directed against plant parts that are lodging the most susceptible stage of the pest so that mortality occurs before its cause's damage to the crop. Also directional application prevents inactivation of the microbes and increases the chances of infecting the target pests (Narayanan, 2007)

Spray Fluid Volume

Studies carried out by Chapman and Ignoffo (1972) indicated that doubling the spray fluid volume of *Ha NPV* proportionately increased the mortality of *H.zea.* Very few studies are been conducted in the requirement of spray fluid volumes in relation to efficacy of microbes in horticultural pest management..

Formulations with Different Additives

Among several factors, that limit the efficacy of insect pathogens in the field, factors such as temperature, humidity, UV radiation and antagonists play an important role. The effect of these factors can be minimized by addition of suitable additives. Formulations play an important role in stabilizing the pathogens under field conditions. The need for improved performance specifications has led to development of a variety of adjuvant for addition to tank mixtures. Additives are added to increase wettability, adhesiveness, stability suspensability, flowability, decrease evapouration, sunlight degradation and to act as gustatory stimulants.

Phago Stimulants

As indicated, for virus and bacteria, the spores are to be ingested in sufficient quantities for infection to pick up. Some of the feeding stimulants like cotton seed oil, water extract of corn kernel, plant extracts, crude vegetable oils soya flour, crude sugar, chick pea flour, soya oil tender coconut have been reported to increase the efficacy of NPV of *H. armigera* and *S. litura.* The addition of feeding stimulants apparently made the virus more palatable to the larvae, resulting in increased ingestion of the virus and thus higher larval mortalities. Addition of various adjuvants such as soya flour (1%) and soya oil (0.01%) [Narayanan and Gopalalakrishnan,1988] molasses 5% + tinopal 0.2% + lamp balck (0.1%) to *Ha NPV* and cotton seed kernel extract, corn oil 1.5%, crude sugar 25%, Tween to *Sl NPV* were reported to increase their efficacy against *H.armigera* and *S.litura* respectively under field conditions (Annoymous, 2005b & 2006b)

Protectants

UV protectants are recommended to reduce the thermal inactivation of microbes against target insect pest. Addition of whitening agent such as robin blue and Tinopal was reported to increase the efficacy of *Ha* NPV on tomato (Narayanan and Gopalakrishnan, 1988) and *Sl* NPV (Anonymous, 2005 b).

Persistence

Persistence denotes the ability of pathogens to remain present in the active state. Environmental factors such as low RH and high temperature causes desiccation of fungal pathogens thereby inactivating them. Formulating fungal pathogens in oil [Prior, et.al., 1988] increases their effects, probably by preventing conidial desiccation, increasing adhesion, spreading the innoculum and interfering with the defense nature of cuticle. These could help in bye passing the first barrier to pathogenesis by fungal spores.

Studies have indicating that vegetable oils and mineral oils such as sunflower oil, coconut oil, neemoil, glycerol, and paraffin are compatible with *Metarhizium anisopliae, V.lecanii* (Ganga visalskhy, et.al. 2005). Addition of oils such as coconut oil, gingely oil was reported to increase the efficacy of *Nomurea rileyi* against *H.armigera* on tomato (Anonymous, 1997), *M.anisopliae* against mango hopper and chilli thrips (pers.observation). Similarly, Hussaini et.al, (2005, a and b) *has* reported that addition of various adjuvants with EPN could prolong their desiccation and thereby increase their effectiveness. However, all additives are not compatible with microbes. Hence, it is essential to determine the compatibility with each species of microbes before taking into formulation stage

Application Methodology

Very less work has been done on application techniques for microbes against insects. Microbes can be applied as dry material, wettable powder, using materials such as kaolin, silica or carriers, dusts etc. similarly, EPN, virus bacteria can also be applied as liquid, WP, granules etc. Studies on the effectiveness of microbes in relation to the insect behavior and crop needs to be standardized for obtaining successful results. In addition, the effect on human beings and other organisms are to be noted. For example dusts are vulnerable to be lost from target areas by wind and dry application can also cause allergic responses in field persons.

Integration of Microbial in IPM

The success of microbes as biological control agents depends on how well the identified species is compatible with other control methods to develop IPM of the particular crop system. Hence it is essential for a potential microbe to be compatible with other chemical that are recommended as plant protection for effective suppression of the pest and disease problems.

In many instances the microbes alone may not be effective. But integration of a sub lethal dosage of the insecticides was reported to give effective control. Combination of Chlorodimecron @0.25kg/ha with Dipel gave better result (Krishnaiah et.al.1981). Similarly when more than a pest complex occurs, since microbes are host specific many a times there isa need to apply insecticides. Narayanan et.al.1975 has reported that dimethoate and DDT are compatible with Bt. Hence combination of these insecticides with the microbe is reported to control DBM as well as aphids *Myzus persicae* that occur simultaneously on cabbage.Spraying of NPV of *Adisura atkinsoni* @250LE/ha + reduced dosage of endosulfan @0.035% was found to be effective in management of pod borers such as *A. atkinsoni, H, armigera* and *Sphaenarches anisodactyhus* (Narayanan, 1985). Ganguli *et al.* (1997) recommended one application of NPV @ 250 LE/ha at the time of pest occurrence followed by spraying of endosulfan 0.07% to protect the crop from *H. armigera.* In general, *Ha* NPV with endosulfan, both at reduced doses is recommended for minimizing the borer damage effectively (Ganguli *et al.*, 1997).

In case of fungal pathogens, combination of fungal suspension of *V.lecanii* with quinalphos at $1/10^{th}$ of the recommended dosage was reported to give effective control of the scale insect *Coccus viridis* on coffee (Eswara moorthy and Jayaraj, 1977). Many of the fungicides such as monocrotophos, dimethoate, phosphamico and fungicides such as Captol, Zineb, Cholothaloni, and Ziram were reported to be safe *N. rileyii*, a potential microbial control agent of lepidopteron pests of horticultural crops. Hence there is a good potential of this fungus being integrated with the IPM strategy of *H.armigera* on tomato. (Gopaplakrishnan and Mohan, 2000). Similarly, combination of the entomopathogen *M. anisoplaie* for the control of mango hopper was found to be compatible with fungicides Sulfex, Bayleton, Karathan that are recommended for disease managemtn during flowering [Ganga visalskhy, pers.obervation].

BIOLOGICAL APPROACH

Strain Selection and Strain Improvement

Pathogen needs to be virulent, specific amenable to culture and able to withstand the environment in which they are used. It is necessary to distinguish and define clearly the characteristics of fungal isolates (which may morphologically indistinguishable) with activity against specific insects and here the techniques of molecular biology are important.

There is great potential for collecting new genotypes from well known genera. Search among the less well known genera may also prove to be valuable. Limited host records usually reflect an absence of insect pathologists rather than real lack of entomo-pathogens and this is especially true for many tropical countries where host records are extremely rare but species diversity is at its highest.

Better genotypes may become available by collecting new material or by improving those already available by conventional approach or biotechnological approaches. Obvious characteristics to select for include virulence, specificity and resistance to adverse environmental conditions and compatible with IPM strategy (Anonymous, 2007-09).

Integration with other Biological Control Agents

The era of 21st century aims at sustainable agriculture through IPM strategy. That is to promote technically sound, economically viable, environmentally degrading and socially acceptable natural resources. More and more importance on pest management involves a sound knowledge of the ecology of key pests. Biological control by microbes forms an important component in this aspect. These pathogens when conserved and augmented would keep some of the pests under check within the economic threshold level. In other cases, the pathogens are cultured in vitro are employed either for short term of control equivalent to chemical control for long term biological control. As said above many a times integration of microbial control agents may be needed for management of different insect species occurring on a crop or integration may also help to attack simultaneously two stages of the same insect for successful control. Hence, the success of microbes as components of BIPM depends on how compatible are they?

The NPV of *H.armigera* was found to be compatible with other biological control agents such as egg parasitoids and B.T. products. Integration of these two agents is reported to bring significant control than application of a single agent. Krishnamurthy *et al.* (1999) suggested the release of *T. pretiosum* (2.5 lakhs/ha) + 2 sprays of *Ha* NPV (250 LE/ha) for the effective management of tomato fruit borer than either alone. Gupta and Rajaram Mohan Babu (1998) found that three releases of *T. pretiosum* + three sprays of Bt @ 1kg/ha were found to be highly effective in reducing the damage caused by *H. armigera* on tomato in Himachal Pradesh. It was reported that Bt products are safe to the egg parasitoid *Trichogramma chilonis*, an egg parasitoid of brinjal borer (Ganga Viasalakshy, 2006). Integration of *T.chilonis* with Bt products is reported to significantly reduce the pest incidence than either one only against brinjal shoot and fruit borer (Ganga Visaskhy, 2004). Devaraj *et al.* (2000) found that the application of a mixture of *Sl* NPV and *Ha* NPV effectively reduced the pest population of *S.litura* (F.) and *H. armigera* thereby increasing the yield. Spraying of Dipel for management of DBM in addition to being suppressing the pest is reported to be safe to other naturally occurring promising biocontrol agents such as larval parasitoids Cotesia *plutella* (Krishnamoorthy *et.al.*2005).

However detailed studies on the compatibility have to be made before integration. Many a times the bioagents could have deleterious effects to each other as reported for the plant antagonists *Trichoderma viridae* and *T.harzianum* against the entomofungl pathogens such as *Metarhizium anisopliae* and *Verticiluum lecanii* (Ganga visalakshy *et.al.*2007).

BIOTECHNOLOGICAL APPROACH

Even though wild type microbes are utilized for more than three decade their wild spread and acceptance still remain at its infancy. The advent of biotechnology provided a new opportunity to make these organism efficient bio-pesticides by using them to deliver a foreign gene to its own gene or deleting some genes which would have a lethal impact or other wise in increasing their efficacy and utilization. The ability to enhance the insecticidal properties of each group of entomo-pathogenic genes is related to the degree of understanding of the biology of pathogenecity and ease of genetic manipulation.

There are a variety of proteins and peptides such as insect neurotoxins that act within the haemocoel that are not active against an insect pests by ingestion or by topical application. Insect virus such as baculovirus provides a delivery system for these toxins into the haemocoel of the insect, where they have access to the site. Such toxins provide a second line of offense in addition to virus itself. Most progress has been made toward since late 1980,s and has resulted in products that now approach the efficacy of chemical insecticides (Harrison and Bryong , 1998).

Bacillus thurirnigiensis has also received considerable attention because of the insecticidal toxins produced by different strains. As these toxins are active after ingestion by a variety of insect pests they have been widely exploited for promotion of transgenic plants and optimization of the bacterium itself as a microbial insecticide.

Optimisation of entomo-pathogenic fungi by genetic engineering is in its infancy because of limited knowledge of molecular and biochemical basis for fungal pathogenesis. Identification of genes involved in pathogenesis is essential for improvement of insecticidal efficacy by genetic manipulation and there are likely to be several broad classes of genes involved with pathogenecity at different stages of fungus –host interaction (St.Lanker, 1993). The host specificity of entomopathogenic fungi is determined in part by the specificity of proteases produced on contact with the host cuticle. One of the goals for future could be to tailor a fungus for control of target pest species by manipulation of protease genes. Site directed mutagenesis of enzymes by manipulation of enzymes involved in pathogenesis could also be used to elucidate the catalytic mechanisms of the enzymes with the potential for enhancement through genetic manipulation.

Engineering of entomopathogenic nematodes could potentially be used to enhance pathogenecity, improve environmental tolerance and alter host range once the genetic bases for these traits have been elucidated (Poinar, 1991). Researchers working on optimization of entomo-pathogenic nematodes at the generic level are able to exploit information gained from the relatively closely related *Caenorhabidtis elegans* which has been extensively studied as a model organism for developmental processes (Zwal.*et.al*. 1993).

IMPLEMENTATION

Microbial Control finds useful Application in Developing Countries against Several Agricultural Contexts

1) Where over use of insecticides have created pesticide resistance so high that need for alternate control becomes necessary. In this context conservation coupled with the reduction in pesticide use may result in effective control which may further be enhanced with augmentation and introduction.
2) Microbial control may be introduced if economical and compatible with existing methods. The main draw back is the host specificity of microbial agents. Hence their effectiveness is lost when introduced in a pest complex situation. Thus the best systems amenable for microbial control are those with one or a few major pests that are economically important.
3) Microbial could work well in organic farming and in protected cultivation as the biotic factors are very congenial for its effectiveness.

MAJOR THRUST AREAS FOR FUTURE RESEARCH ON MICROBIAL AGENTS OF INSECT PESTS OF HORTICULTURAL IMPORTANCE

1) Build up a wide base of microbe repository with rich biodiversity characters for suitable strain isolation, and improvement.
2) characterization by molecular methods
3) Tritropic interaction studies on the host plant- target pest and microbes

4) Develop conservation strategies for microbes that form potential factors in regulating pest populations.

5) Genotype improvement of the target microbes with desired characters for increased effectiveness

6) Develop cost effective product for commercialization

7) Studies on formulation and application technology to increase the shelf life, viability and efficacy

8) Integration of other control methods and popularization of microbial control by demonstrating economic and ecological benefits of microbial control

9) Community involvement approach for production, management and providing market linkages for price incentives for the crop produced by bio-pesticides.

CONCLUSION

The biological potential of microbes in management of insect pest in horticulture is undoubted. Alternate methods of pest control have become an integral part of IPM programs as the concern over the ill effects of conventional chemicals has increased in the pats two decades. As these compounds are removed from the market place there is an on going search for effective products that have minimal impact on beneficial organisms. Microbial pesticides are presently experiencing a renaissance as regulatory constraints are resolved. Their success however will ultimately depend on the willingness of consumers to employ new methods of pest control and integrate microbial into IPM programmes.

REFERENCES

[1] Anonymous (2007 a). Annual report, IIHR.

[2] Annoymous, (2005) Annual report, IIHR.

[3] Annoymous, (2006 a), Annual report, IIHR.

[4] Annoymous, (2006 b), Annual report, IIHR.

[5] Annoymous, (2007 b). Annual report, IIHR.

[6] Anonymous, (2008) Annual report, IIHR.

[7] Anonymous, (1997).Annual report, IIHR, Bangalore ,

[8] Anonymous, (2006 c). Development of Bio-control based management strategy for brinjal shoot and fruit borer Leucinodes orbonalis, Final report submitted to Government of Karnataka.

[9] Asokan, R. & Mohan, K.S. (1996). Use of *Bacillus thuringiensis* Berliner for the management of fruit and vegetable crop pests – An appraisal. *Agricultural Reviews*, 41: 238-248.

[10] Asokan, R, Mohan, K.S. & Gopalakrishnan, C. (1996). Effect of commercial formulation of *Bacillus thuringiensis* Berliner on yield of cabbage. *Insect envt.,2:* 58-59

[11] Atwal, A.S. & Singh, K. (1969). Comparative efficacy of *Bacillus thuringiensis* Berliner and some contact insecticides against *Pieris brassicae* L. and Plutella maculipennis (Curtis) on cauliflower. *Indian J. Ent.,* 31: 361-63.

[12] Battu, G.S. (1995). A note on the nuclear polyhedrosis of the cabbage caterpillar *Pieris brassicae*. *Annals of Plant Prot. Sci.*, 3: 164.

[13] Battu, G.S., Singh, H., & Singh, L., (1989). Nuclear polyhedrosis virus infection of *Hellula undalis* Fab. *J. Insect Sci.*, 2: 60-61.

[14] Burghes, H.D. (Ed) (1981). *Microbial control of pest and plant Diseases.* Academic press, London

[15] Chapman & Ignoffo (1972). Influence of rate and spray volume of a nuclear polyhedrosis virus on control of *Heliothis armigera* in cotton .*J. of Invertebrate Pathol.* 20:183-186

[16] Datta, S.(1992) *Biotechnology in integrated pest management* In Farming system and integrated pest management (J.P. Vermand A.verma eds.) Malhothra publ.house, N.Delhi., pp57-61

[17] Datta, P. K. & Sharma, D. (1997). Effect of *Bacillus thuringiensis* Berliner in combination with chemical insecticides against *Spodoptera litura* F.in cabbage.*Crop Research,*14: 481-486

[18] Devaraj, R., Rabindra, R.J., Kennedy, J.S. & Sathiah, N. (2000). Combined application of Sl NPV and Ha NPV against *Spodoptera litura* (F.) and *Helicoverpa armigera* (Hub.) populations on tomato. Paper presented in *National Symposium on Microbials in insect pest management,* Loyola College, Chennai, Feb 24-25, 2000.

[19] Dhandapani N. Jonathan, R. & Kumarasami, T. (1992*).* Epizootic occurrence of nuclear polyhedrosis viruses on *Spodopetra litura* F. *Curr. sci.,*51: 793-794

[20] Dhandapani, N. & Jayaraj, S. (1989). Efficacy of nuclear polyhedrosis virus formulation for the control of *Spodoptera litura* (Fab.) on chillies. *J. Biol. Control,* 3: 47-49.

[21] Easwaramoorthy, S. & Jayaraj, S. (1977). Effect of certain insecticides and fungicides on the growth of the coffee bug Fungus *Cephalosporium leacnii* Zimmerman. *Madras Agricultural Journal,* 64: 243-246

[22] Ganga Visalakshy P. N. & Krishnamoorthy, A. (2006). Improving the efficiency of *Trichogramma chilonis* Ishii in the management of brinjal borer by integrating with botanicals and biopesticides – Paper presented in *National Symposium on Improving Input Use Efficiency in Horticulture,* August 9-11, IIHR, and Bangalore.

[23] Ganga Visalakshy, A, Manoj Kumar & Krishnamoorthy, A. (2004). Epizootics of a Fungal Pathogen, *Verticillium lecanii* Zimmermann on *Thrips palmi* Karny. *Insect Environment.* 10 (3): 134-135

[24] Ganga Visalakshy, Krishnamoorthy, A. & .Hussaini S. S. (2009). *Field efficacy of entomopathogenic nematode Steinernema carpocapsae (Weiser, 1955) against brinjal shoot and fruit borer, Leucinodes orbonalis Guenee. Nem. Mediterranea (in press)*

[25] Ganga Visalakshy, P.N., Krishnamoorthy, A. and Manoj Kumar, A. (2007). Compatibility between the plant antagonists *Trichoderma harzianum* Rifai, *T. viride* Pers.:Fr. and potential entomopathogenic fungi of horticultural crop pest. *Entomon,*31(2):129-132

[26] Ganga Visalakshy, P.N. Krishnamoorthy, A & Manoj Kumar, A. (2005). Effects of Plant Oils and Adhesive Stickers on the Mycelia Growth and Conidiation of *Verticillium lecanii*, a potential entomopathogen. *Phytoparasitica,* 33(4): 367-369

[27] Ganga Visalakshy, P. N. & Krishnamoorthy, A. (2004). Biological Control based IPM of egg plant shoot and fruit borer, paper presented in the *National Symposium on Recent*

advances in integrated management of Brinjal shoot and fruit borer at IIVR, Varanasi from 3-10-05 to 4-10-05.

[28] Ganga Visalskhy, P.N., A, Manoj Kumar, A. Krishnamoorthy, A & Mani, M. (2007 b). Pathogenecity of entomopathogenic fungus Veticillum lecanii to Scirtothrips dorsalis Hood. *In National Sym. On biological control of sucking pests*, Bangalore, 26-5-06 to 27-6-06.

[29] Ganga Visalskhy, P.N.,Krishnammorthy, A, Anurrop, C.P. & Aparna, K.(2009)Reduction of mango inflorescence hopper, *Idioscopus nitidulus (Walker) population following field application of the fungus Metarhizium* anisopliae, presented in *International conference eon biopestiicdes- stake holders perspectives* held at Delhi from 26- 4-09 to 30-4-09

[30] Ganguli, R.N., Singh, V.V., Dixit, S.A. & Kowshik, U.K. (1997). Efficacy of NPV (nuclear polyhedrosis virus) and endosulfan against tomato fruit borer, *Heliothis armigera. Curr. Res.,* 26: 210-212.

[31] Gelernter, (1997). Resistance to microbial insecticides, In: microbial insecticides novelty or necessary? BCPC symposium proceedings No. 68: 201-212

[32] Gopalakrishnan, C & Ganga Visalakshy, P.N. (2005). Field efficacy of commercial formulations of *Bacillus thuringiensis var krustaki* against *Papillo demoleus* L on citrus. *Entomon,*30 (1): 93-95

[33] Gopalakrishnan, C. & Narayanan, K. (1989). *Epizootiology of Nomuraea rileyi (Fallow) Samson in field populations of Helicoverpa (=Heliothis) armigera Hubner (Noctuidae: Lepidoptera) .Curr. Sci.,* 57: 867-868.

[34] Gopaplakrishanan & Mohan, (2000). Effect of certain insecticides and fungicides on the conidial germination of *Nomurea rileyi. Entomon,* 25: 217-223.

[35] Gupta, Rajaram & Mohan Babu (1998) *Management of Helicoverpa armigera on tomato with Trichogramma pretiosum and Bacillus thuringiensis var. kurstaki.* Proc. I *Natl. Symp. Pest Mgmnt. Hort. Crops,* Oct, 15-17, 1997, Bangalore, pp. 75-80.

[36] Harper, (1987). Applied Entomology: microbial control of insects in epizootology of insect diseases (J.R.Fuxa and Tanada.Y.eds.)pp 473-496.

[37] Harrison R.L. & Bryong C. Bonning (1998). Genetic engneering of biocontol agents for insect. In Biological and biotechnological control of insect pests. Eds. Jack E. Rechecking and Nancy A. Rechcigl. Lewis publishers, Washington.

[38] Hussaini & Singh.(1998) Entomophilic nematodes for control of insect pests In *Biological suppression of plant diseases, phytoparasitic nematodes and weeds* Eds. Singh ,S.P. and Hussaini, publ. by Project Directorate of Biological Control, Bangalore.

[39] Hussaini, S.S., Nagesh, M. Rajeshwari, R. & Shahnaz fathima.(2005.b) Effect of adjuvants on survival and pathogenecityof some indigenous isolates of EPN. *Ind.J.of plant protection,*32(2): 111-114

[40] Hussaini, S.S., Nagesh, M. & Shakeela, V. (2005).A effect of antidessicants on survival and pathogenecity of some indigenous isolates of EPN against *Plutella xyllostella. Annals of plant protection sciences,*13(1): 179-186

[41] Krishnaiah, K., Jaganmohan, N. & Prasad, V.G. (1981). Efficacy of *Bacillus thuringiensis* Ber. for the control of lepidopterous pests of vegetable crops. *Entomon,* 6: 87-93.

[42] Krishnamoorthy, A., Gopalakrishnan, C. & Mani, M. (1999). Integration of egg parasitoid and Ha NPV for the control of Helicoverpa armigera (Hubner) on Tomato.

Paper presented in the *International seminar on Integrated Pest Management,* IICT, Hyderabad, Oct. 8-9, 1999.

[43] Krishnamoorthy, A. & Ganga Visalakshy, 2007. V[th] *QRT report of AICRP on Biological control of crop pests and weeds of IIHR,* Bangalore from 2002-2007.

[44] Krishnamoorthy, A., Gopalakrishnan, C. & Mani, M. (1999). *Integration of egg parasitoid and Ha NPV for the control of Helicoverpa armigera (Hubner) on tomato. Paper presented in the International seminar on Integrated Pest Management,* IICT, Hyderabad, Oct. 8-9, 1999.

[45] Krishnamurthy, P.N, Krishnamoorthy, A & Ganga visalakshy. P.N. (2005) IPM on cabbage and cauliflower, *Extn folder IIHR, No. 6*

[46] Lawrence & Lacey. (1997).Manual of techniques in insect pathology, Academic press, NY.

[47] Mani, M., Gopalakrishnan, C. & Krishnamoorthy, A. (2000). Natural parasitism on the pomegranate hairy caterpillar Trabala vishnou Lefevere (Lepidoptera: Lasiocampidae) in Karnataka. *Entomon,* 25: 241-243

[48] Mani, M., Gopalakrishnan, C. & Krishnamoorthy, A. (2001). Natural enemies of ber hairy caterpillar Thiacidas postica Walker (Lepidoptera: Noctuidae) in Karnataka. *Entomon,* 26: 313-315.

[49] Mani, M. Krishnamoorthy, A & Gopala Krishnan, C. (2005). Biological control of lepidopterous pests of horticultural crops in India – a review. *Agric.rev.*26: 39-49

[50] Mohan, K.S., Asokan, R. & Gopalakrishnan, C. (1996). Isolation and field application of a nuclear polyhedrosis virus for the control fruit borer Helicoverpa armigera (Hubner) on tomato. *Pest Management in Horticultural Ecosystems,* 2**:** 1-8.

[51] Narayanan & Gopalakrishnan. (1988). Microbial control of Heliothis armigera. Tech.bull.6, IIHR Bangalore.

[52] Narayanan, K. & Gopalakrishnan, C. (1987). Effect of nuclear polyhedrosis virus for the control of Heliothis armigera on tomato. *International Heliothis Biological Control Newsletter,* 6: 10.

[53] Narayanan.K. (2007). Biological and Biotechnological approaches for effective utilization of insect virus for the management of insect pests and beyond *In Biotechnology and insect pest management eds.* Dr.S Ignachimuthu and Dr. S. Jayaraj. Elite publishing house, New Delhi.

[54] Narayanan. K. & Veena kumari.K (2002).. Report of occurrence of nuclear polyhedrosis virus on gherkin caterpillar, Diaphania indica(Saunders) (Lepidoptera:Pyralidae): In Proceedings of the symposium on *Biological control of Lepidopteran pests, July,17-18,*

[55] Patel, R. Singh.R & Patel, P.(1968). Nuclear polyhedrosis virus in gram pod borer *Heloithis armigera J.Econ. Entomology* ,61: 191-193

[56] Phadke, C.H. & Rao, V.G. (1978). Natural out break of the muscardine fungus Nomurea rileyi (FABR) Samson on leaf eating caterpillar, *Spodoptera exiqua* Hb.in Maharastra.*Curr. science,* 47 : 476

[57] Poinar, G.O. *Genetic engineering of nematodes for pest control in biotechnology for biological control of pests and vectors,* Marmarosch, K.Ed.CRC press, Boca Raton, FL, 77.

[58] Prior, C. Jollands, P. & Li Patourel, G. (1988). Infectivity of oil and water formulations of Beauveria bassiana (Deutromycetes) to coca weevil pest Panotorhytes plutus (Coleopteran- Curculionidae) J.of invertebrate pathology, 52: 66-72

[59] Rabindra, R. J. Geetha, N. Renuka,S. & Regupathy, A.(1998). Occurrence of a granulosis from two populations of Plutella xylostella L) in India .the management of DBM and other cruciferous pets. *Proc. of third international workshop Kualalumpur,* Malayasia29, Oct. to 1ˢᵗ Nov. 19969 Eds. A Sivapragasam.W.H. Loke A. k. Hussan and G.S.Lim) pp 113-115 Malaysian plant Protection Society,1996

[60] Ramakrishnan & Tewari. (1969). Polyhedrosis of *Prodenia litura* Fabr.*(Noctuidea) Indian J. of Entomol.*31: 191-192

[61] Satapathy, C. R. & Panda, S.K. (1997). Effect of commercial Bt formulations against fruit borers of okra. *Insect Environment, 3, 2,* 54.

[62] Sharma, S.S. & Chaudhary, S.D. (1994). Bioefficacy of Bacillus thuringiensis and Neem seed kernel extract against Spodoptera litura invading cauliflower. *Haryana J. Horti. Sci.,* 23: 261-262.

[63] Sitandanum, R.Varatharajan, Chandish R.ballal, & Ganga visalskhy. (2007). Research status and scope for biological control of sucking pests in India: Case study of thrips. *J.of Biol.Control., 1-20*

[64] Sivaprakasam, N. (1998). Control of tomato fruit borer Helicoverpa armigera by nuclear polyhedrosis virus in comparison with other treatments. *Indian J. agric. Sci.,* 68: 1-2.

[65] St.Leger, R. (1993). Biology and mechanisms of insect cuticle invasion by deuteromycete fungal pathogens *Iin Parasites and pathogens of insects,* Beckage, N.E.Thompson, S.N. federici, B.A.Academic press, New York, 211.

[66] Tripathi, S.R. & Singh, A.K. (1991*). Some observations on population dynamics of brinjal borer,* Leucinodes orbonalis (Guen.) (Lepidoptera: Pyralidae). Annals of Entomology, 9: 15-24.

[67] Zwal, R.R. Brocks, A.Van –Meurs, Groenene, J.T. & Palsterk, R. H. Target selected gene inactivation in Caenorhabditis elegans by using a frozen transport insertion mutant Bank.*Proc.Natl. acad.Science,* USA, 90, 7431.

In: Recent Trends in Biotechnology and Microbiology
Editors: Rajarshi Kumar Gaur et al., pp. 25-42

ISBN: 978-1-60876-666-6
2010 Nova Science Publishers, Inc.

Chapter 3

GENETIC ENGINEERING FOR VIRAL RESISTANCE IN VEGETABLE CROPS

P. Sharma[1], Richa Raizada[2], T. Ogawa[3] & M. Ikegami[3]

[1] Division of Crop Improvement, Directorate of Wheat Research, Karnal 132001, India
[2] Mody Institute of Technology & Science, Rajasthan, India
[3] Department of Life Science, Graduate School of Agricultural Science, Tohoku University, 1-1 Tsutsumidori-Amamiyamachi, Aoba-ku, Sendai, 981-8555, Japan

ABSTRACT

Plant viruses have a dramatic negative impact on agricultural crop production including vegetable crops throughout the world & consequently, plant pathologists & agronomists have devoted considerable effort toward controlling virus diseases during this century. Prior to advent of genetic engineering, traditional plant breeding methodology was successfully applied to develop resistance to viruses & conventional methods for control of plant virus disease like control of vector population using insecticides, use of virus-free propagating material, appropriate cultural practices, cross protection & use of resistant cultivars. However, each of the above methods has its own drawback. In recent years, the advancements in plant molecular virology have enhanced our underst&ing of viral genome organizations & gene functions. Moreover, genetic engineering of plants for virus resistance has recently provided promising additional strategies for control of vegetable viral disease. At present, the most promising of these has been the expression of coat-protein coding sequences in plants transformed with a coat protein gene. Other potential methods include the expression of anti-sense viral transcripts in transgenic plants, the expression of viral satellite RNAs, RNAs with endo-ribonuclease activity, antiviral antibody genes, & Ribosome inactivating genes in plants.

INTRODUCTION

The control of vegetable diseases caused by viruses is very important for maximizing yield but not direct control measures are presently available to farmers. A few conventional methods include control of vectors that transmit them by use of insecticides, use of virus free propagating materials & cultural practices such as sanitation, crop rotation, programmed sowing & eradication. Conventional breeding programs to develop resistant vegetable varieties are effective but expensive & often in addition to the desired character some of the undesiredable characters may be transferred. Also, the resistance can be circumvented by the variation. Disease resistance is one of the most environmentally single safe methods for controlling plant diseases. Unfortunately in many plant virus systems resistance is not available & cannot be obtained by traditional plant breeding methods. The advances of the past decade in plant transformation & genetic engineering of virus resistance have provided some of the most dramatic developments in our ability to increase genetic resistance to viral diseases. Infect, engineered resistance is one of the first successful demonstration of the introduction of any agriculturally useful trait into plants (Gasser & Fraley, 1989). There are number of ways in which resistance could potentially be achieved by using plant transformation & regeneration, several of these methods have been explored in considerable details. For examples plant transformation might allow the natural virus resistance genes present in one plant species to be introduced into another species. However, the problems in identifying & characterizing such resistance genes limit the use of this approach. Other approaches have included introducing sequenced encoding antiviral proteins such as ribosomal inacting proteins (Baulcombe, 1994, Wilson & Davis, 1994). One of the major contributing factors to the early, rapid & wide development of genetically engineered virus resistance is the source of gene those are used. The practical significance of our ability to use viral gene is that they generally are easy to clone. The use of such genes, however, is only possible with the concurrent development of appropriate plant transformation technologies. To date, many vegetable viruses have been characterized at the molecular level & their genome organization & gene function elucidated. Thus fundamental information needed to employ genetic engineered resistance strategies i.e. presently available in much virus system. Precise interactions between plant & virus components are necessary for virus infection & replication as they rely on host's life cycles. These interactions comprise a number of different processes that allow the virus to uncoat its genetic information, express its gene products, replication, cell-to-cell movement & spread long distance through out the plant. These processes are important for infection & the development of disease. Interference of any of these precise interactions could inhibit the infection process & curtail disease development. Thus information obtained at molecular level cold used to develop & tailor strategies to disrupt plant virus interactions that are important in the infection process.

A. Pathogen Derived Resistance

The concept of pathogen-derived resistance (PDR) was proposed by Sanford & Johnston (1985) & suggested that the possibility of engineering resistance by transforming a susceptible plant with genes derived from the pathogen itself. This form of resistance, which they termed " parasite derived resistance" (subsequently termed as pathogen derived resistance') was envisaged to operate through the expression of the viral gene product at either an wrong time, wrong amounts, or in an inappropriate form during the infection cycle, thereby perturbing the ability of the pathogen to sustain an infection. This proposition was a natural corollary to the phenomenon of cross protection observed in plants infected with mild strain of a virus to the more severe strain. The first illustration that PDR is indeed a viable way of producing virus resistance plants was provided by the experiments of Powell –Able *et al* (1986), who demonstrated that tobacco plants transgenic for, & expressing, the TMV coat protein was resistant to infection with the virus. This discovery opened a new field of research in plant science.

Applications of PDR

Plant viruses differ considerably in their morphology & in the genetic material used to encode the virus genes. These genomes include single or double str&ed DNA, double str&ed RNA or single str&ed RNA of either plus or minus sense. They are either monopartite (with one genome) or multipartite (with divided gemones). They replicate & propagate by various methods but the common feature in the life cycle of all these is the transcription of mRNAs for translation of structural & non structural proteins that are required to fulfill the viral life cycle (Scholthof *et al.*, 1993).

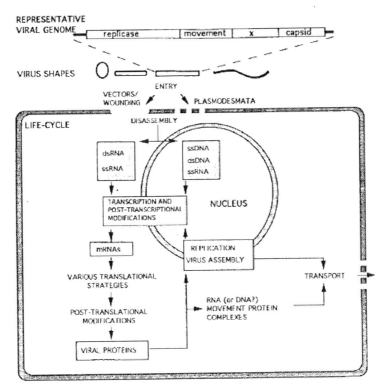

Figure 1. Schematic representation of a plant virus life cycle & replication strategies in a plant cell.

Despite differences in their replication strategies, all plant viruses have broadly similar steps in their life cycle (Figure 1). Upon gaining entry into the host, via fungi, insects, mites, nematodes etc. through necrotic lesions, or through injury caused by abrasives used during mechanical inoculations, the virus particles partially disassemble, to expose the viral DNA or

RNA to the cellular milieu (Verduin, 1992). DNA viruses generally enter the nucleus & utilize host enzymes to produce mRNAs for translation while viruses having RNA as genetic material undergo translation to produce the virus specific proteins necessary for propagating themselves. A crucial process in the infection by most plus sense RNA viruses is the production of replicase protein(s) that along with the cellular machinery produce progeny by replicating the parental genomes. The progeny viral RNA moves from cell to cell via the plasmodesmata as a complex with viral protein (movement protein) that aids in the movement or for optimum long distance, in association with a functionally active coat protein (Citovsky & Zambryski, 1991).

It is theoretically possible to block the viral life cycle at any stages ie uncoating, translation, replication, & or movement. The main aim of generating transgenic plants resistant to virus infection is to express a portion of the viral genome, with or without expression of an encoded protein that will inhibit some of the above-mentioned steps in the multiplication cycle.

1. Coat Protein Mediated Resistance (CPMR)

Coat protein mediated resistance in plants was demonstrated for the first time in TMV (Powell-Abel et al., 1986). Coat protein resistance refers to resistance caused by the expression of a virus coat protein (CP) gene in transgenic plants (Beachy et al., 1990). Accumulation of the coat protein confers resistance to infection & / or disease development by the virus from which the coat protein gene was derived & by related viruses. Resistance is not due to somaclonal variation caused by the transformation & is stably inherited. The majority of examples of engineered resistance to date have used viral coat protein genes. Transgenic tobacco plants constitutively expressing the TMV CP gene were more resistant to infection by TMV then control non-transgenic plants. Since that time, CP mediated resistance conferred by the expression of CP gene for different plant virus groups has been accomplished. The strategy known as coat protein mediated resistance (CPMR) is based on the introduction into the genome of a plant a chimeric gene containing viral coat protein sequences. Although examples were predominantly applied to viruses with genomes of a positive (messenger) sense ssRNA & a single type of protein in their capsid, CPMR has now successfully demonstrated in a variety of host-virus combinations that includes viruses infecting monocot crops (Hayalawa et al., 1992, Murry et al., 1993) & viruses with a negative str& RNA (Beachy et al., 1990; Pang et al., 1992 & Wilson , 1993). Recently, CPMR phenomenon has been also reported successful in the case of DNA viruses (Kunik et al., 1997). CP gene is the most widely used transgene to generate transgenic plants. CP gene from at least 37 RNA/DNA viruses belongs to 17 groups has been used to compare resistance into many different plant species (Verma et al., 2002).

Coat protein mediated resistance resulted when the CP gene from U1 strain of TMV was cloned into a plant shuttle vector under the control of the CaMV 35 S promoter & nopaline synthase terminator. The CP was introduced into a tobacco cultivar that is systemic host for TMV infection. A high level of coat protein expression was detected in some of the transgenic tobacco lines. When transgenic plants were inculcated with TMV, CP (+) plants showed later development of disease symptoms than did CP (-) plants, & many plants escaped infection or did not develop symptoms. An absence of disease symptoms indicated a lack of viral replication (Cassalls et al., 1977). When CP(+) plants inoculated with U1 strain of TMV, the inoculated leaves contained less than 30% of the virus found in inoculated leaves of control plants (Nelson *et al.*, 1987).

Mechanism of CPMR

The mechanism of resistance induced by CP gene is either the protein encoded by the transgene or by the transcript of the transgene or both. CP gene induced resistance is protein mediated when a single copy of transgene is inserted. Resistance so express is of moderate level against a broad range of reluctant viruses & influenced by the level of CP expressed in transgenic plants. Transgene under goes transcription & translation resulting in high levels of protein. The protein inhibits disassembly of the infecting virus & forces the assembly. Disassembly equilibrium towards assembly. Importantly, experimental results indicate that the mechanism varies among viruses. Perhaps this is not unexpected given the fact that, depending on the virus, coat protein can be involved in an array of functions throughout the virus life cycle including encapsidation, cell-to-cell & long distance movement, vector mediated transmission & regulation of replication or translation.

Coat protein mediated resistance in many systems is generally limited to protection against the homologous virus or strains closely related to the integrated coat protein gene. This implies that the coat protein gene of each virus for which resistance is desired must be integrated into the genome of the host in order to obtain broad-spectrum resistance. Even heterologous resistance as also been reported in several potyviruses systems. Transgenic tobacco expressing soybean mosaic virus coat protein showed resistance to TEV & PVY (Lawson *et al.*, 1990) is one of the examples. Namba *et al.* (1992) have reported that transgenic tobacco plant expressing CP of watermelon mosaic virus II (WMV-II) or Zucchini yellow mosaic virus are resistant to varying degrees to at least six different potyviruses. Heterologous resistance is generally less then the resistance conferred against the homologous virus. It has also been reported in transgenic tobacco expressing cucumovirus, CMV strain C, coat protein (Quemada *et al.*, 1991). However two distinct mechanisms seems to be involved, one of the early steps in the replication of viruses is the release of nucleic acid from the virion. The released portion of nucleic acid (mRNA) binds with ribosomes & thereafter the co-translation disassembly leads to the initiation of infection process. Interference in movement of virus in coat protein expressing plant is another way by which symptoms could be delayed. It has also been suggested that TMV coat protein may play a role in regulating the replication of TMV-RNA (Osbourn *et al.*, 1989). Earlier studies in bacteriophages have also shown that replication of several RNA containing phages & expression is affected by their coat protein.

Experimental evidence indicate that possibly the coat protein gene expression interferes with an early event in infection at least in some cases i.e. uncoating. The resistance in some instances is largely overcome by inoculation with viral RNA rather than virons (Loesh-Fries *et al.*, 1987; Nelson *et al.*, 1987; Van Dun *et al.*, 1988). Conversely, in transgenic plants expressing PVX coat protein, the resistance was not broken by challenge inoculation with viral RNA (Hemenneay *et al.*, 1988).

In the re-encapsidation theory, endogenous CP subunits prevent viral uncoating by binding the viral RNA as soon as viral CP subunits are removed. Viral disassembly will be absent or delayed if the levels of CP in the cell are above a minimum level. This theory is consistent with the observation that plants expressing higher levels of CP are more resistant to viral infection than are plants expressing lower levels. The specificity of CP mediated is also explained. If the CP cannot bind to the RNA of the infecting virus, no resistance is possible. Re-encapsidaton was proposed by Dezoeten *et al* (1975) to explain classical cross protection, but it is equally applicable to CP mediated resistance.

In the second theory, endogenous CP binds to putative cellular receptors for viral entry or disassembly. The prior binding of CP to the cellular prevents the incoming virus from entering to the cell or forming disassembly complexes. This explains relatively low (in some cases) levels of CP which can provide viral protection in transgenic plants. In this case the specificity of CP mediated protection depends on the specificity of the viral receptors. In field protection to PVY, a plus sense RNA virus, occur in transgenic plants producing undetectable levels of PVY coat protein (Kanieneski *et al.*, 1990). This important finding might be applied to a number of other vegetable crops because PVY & other potyviruses infect a broad range of plant species & cause yield reductions on economically important crops. CPMR has also been shown to greatly reduce the titer of PLRV in plants transformed with the PLRV coat protein gene. In this case, transgenic plants accumulated PRLV coat protein transcripts, but the PLRV coat protein was not detected (Kawchuk *et al.*, 1991). Lindbo & Dougherty (1992) have postulated that protection sometimes results from coat protein mRNA accumulation & is independent of a requirement for coat protein expression per se.

Table 1. Plant species & viruses for which CPMR has been demonstrated.

Plant	Viruses	Reference
Cucumber	CMV	Gonsalves *et al* (1991), Van Dun *et al* (1987)
Potato	PLRV	Kanieneski *et al* (1994), Presting *et al* (1995), Grahama *et al* (1997)
	PVX	Hoekema *et al* (1989), Lawson *et al* (1993), Kanienes *et al* (1989), Jongedijk *et al* (1993)
	PVY	Kanienes *et al* (1989), Lawson *et al* (1993), Smith *et al* (1995), Hefferon *et al* (1997)
	PVS	Mackenzie *et al* (1991)
	PVM	Mackenzie *et al* (1994)
Soybean	BPMV	Die *et al* (1996)
Squash	WMV	Clough & Hann (1995)
	ZYMV	Triceli *et al* (1995), Funchs & Gonsalves (1995)
	SqMV	Pang *et al*, 2000
Pea	PEMV	Chowria *et al*, 1998
Sugerbeet	BNYVV	Nelson *et al* (1988)
Tomato	TMV	S&ers *et al* (1992)
	CMV	Fuchs *et al* (1996), Provvidenti & Gonslves (1995)
	AIMV	Tumer *et al* (1987)
	ToMV	S&ers *et al* (1992)
	TSWV	Gielen *et al* (1991), De Haon *et al* (1992), Mackenzie & Ellis (1992), Kim *et al* (1994)
	TYLCV	Kunik *et al* (1994)

Tobacco plants expressing a PVY N cDNA that included the 3' end of the Nib polymerase gene the coat protein coding sequence & most of the viral untranslatable 3' region were resistant to several strains of PVY. This resistance was not overcome by mechanical or aphid or graft mediated inoculations. It was also shown that the resistance to PVY was maintained even when the plants were co-inoculated with other potato potyvirus, PVY & PVA. This is relevant because in field situations it was common to find mixed infection & most of the research on genetic engineering resistance usually focuses only on a single virus. Also interesting was the observation that the PVY CP was not detectable in transgenic potato or tobacco inoculated plants.

When the plants were inoculated with a heterologous potyvirus, high accumulation of PVY in CP was observed. Since no AVG was engineered in front of the CP cloning sequence, it is possible that only in the presence of heterologous virus the peptide is processed or stabilized. In potyviruses, the origin of assembly is likely to be in the 5' region of the viral genome (Sit et al., 1994) so that a CP interaction would have the potential to suppress translation of the viral RNA dependent RNA polymerase (RdRp) that is encoded in the 5' most open reading frame (ORF). However it is also possible that CPMR inhibits the cell to cell movement of PVX, for which CP is an essential cofactor (Chapman et al., 1992).

Since alteration in coat protein can influence symptom development & virus spread through the plant, clearly there are pits of interaction between the CP & the host that may serve as points of interference in the infection process. Otherwise in most of the symptoms in which coat protein mediated resistance has been reported have directed against plus sense RNA viruses with a single capsid protein. However, plants expressing the nucleo-capsid protein of the antisense tomato spotted wilt virus (TSWV) have also been shown to be resistant to challenge inoculations (Gielen et al., 1991; Mackenzie & Ellis, 1992; Pang et al, 1992). TSWV particles are membrane enveloped, spherical & composed of at least three structural proteins including the nucleo-protien & two membrane associated glycoprotiens. de Haan et al (1992) found TSWV nucleo-protien protection to be independent of protein expression & thus may be the result of an RNA-mediated protection to be independent of protein effective against inoculations using viruleferous thrips. Additional studies have shown this transgene nucleoprotien resistance to be heterologous against a number of different isolates of the same TSWV serogroup as well as one isolate from a deferent serogroup (Oanf et al., 1992). In field conditions CPMR probably will not work for nematode transmitted viruses (Pleog et al., 1993).

2. RNA Mediated Resistance

Since viruses are essentially packed in nucleic acid that serve as templates for replication, translation & in some cases transcription (DNA viruses). Though it is not surprising that nucleic acids can be used to interfere with normal virus life cycles. The first clue on the molecular basis of RNA mediated resistance was sown by Lindbo et al (1993). RNA mediated resistance approaches have utilized both sense & anti-sense RNAs targeting both coding & non coding portions of the genome. This RNA (or DNA) mediated resistance could operate, if the transgenic nucleic acids act as a decoy molecule. This decoy would compete with the viral genome to redirect host or viral encoded proteins into interactions that would be nonproductive for replication or the spread of the virus in the infected plant. This type of competitive inhibition could operate when the transgenic specified a defective-interfering

(DI) RNA, encompassed a cis-acting element in the viral genome or was derived from a satellite RNA.

Another approach to RNA mediated resistance is to introduce genes that encode the complementary or antisense RNA str&.

Antisense-RNA Mediated Resistance

The anti-sense technology is based on the capability of complementary nucleic acids to form double str&ed helicals. Any single str&ed nucleic acid present in a living cell is therefore susceptible to bind base pairing to a nucleic acid of complementary (antisense) polarity. Specific binding of anti-sense nucleic acid results in interference with biological function of the sense nucleic acid either by arresting the sense nucleic acid or by induced degradation of double str&ed complex. In a view of the efficiency in down regulation gene expression by mRNA directed anti-sense RNAs, it could be expected that replication of plant viruses would also be successfully suppressed by antisense RNAs directed against the pathogenic viral RNA. In general low efficiency of antisense RNA as an antiviral agent against RNA virus's contents with the effectiveness of mRNA directed antisense RNA in gene suppression. One reason that accounts this difference is that antisense RNA & the viral target RNA are synthesized in different cellular compartments, whereas the stably transformed antisense gene is transcribed in the nucleus, the viral target RNA is synthesized by the viral replicase in cytoplasm. Only a small fraction of the nuclear antisense RNA is actually transported in intact form to the site of virus replication. So duplex formation is a bimolecular reaction, & requires that the antisense RNA is present in at least stoichiometric amounts at the site of virus replication to be effective.

Studies involving CP mediated resistance against PVY & PLRV have also identified transgenic lines that expressed little or no detectable lines of coat protein, yet were still resistant (Lawson et al., 1990 & Kawchuk et al., 1991). Thus, resistance conferred in the coat protein systems may actually be due to an RNA sense mediated mechanism. Transgene constructs containing the complete or partial ORF of the PVX 165 KDa replicase protein did confer resistance (Braun et al., 1992). Protein from there PVX constructs was not detected, resistance due to the interfering properties of sense RNA may be a factor in this phenomenon.

Several proteins of plant viral genomes have been targeted for their potential in conferring RNA mediated resistance. Expression of antisense sequence of the AC1 gene involved I replication of the TGMV geminivirus lead to decreased TGMV infection (Dry et al., 1991). It is likely that resistance from a transgenically expressed 3'DI-RNA of turnip yellow mosaic virus is conferred through direct competitive inhibition with the viral genome (Zaccomer et al., 1993). The majority of antisense & sense defective constructs targeted to date have utilized coat protein sequences. In many cases the level of protection conferred by antisense sequences was considerably lower then that conferred by the corresponding coding sequence of the CP gene of CMV (Cuzzo et al., 1988), PVx (Hemenneay et al., 1988), ZYMV (Pang et al., 1993) & PVY (Farienelli et al., 1993). However in other examples both sense & antisense str& of RNA appeared to be comparably effective against PLRV & BYMV (Kawchuk et al., 1991; Hmmond & Kanio, 1993). Similarlily better levels of protection were observed for plants expressing translation defective versus translatable constructs of CP genes of TSWV (de Haan et al., 1992) & PVY (Farienelli et al., 1993; V&er Vlugt et al., 1992).

Cuzzo *et al.* (1998) showed that plants expressing antisense CMV RNA were resistant to infection. In such plants movement of virus was slower & accumulation was less than in control plants. Effect was however only detected when the inoculum concentration was low. Similar observations have been made by Hemenneay *et al.* (1988) & Powell *et al.* (1989) with antisnse PVX RNA & TMV RNA, respectively. However expression of TRSV RNA did not enhance the resistance of plants to TRSV infection (Baulcombe *et al.* 1987; Angenent *et al.*, 1990). RNA mediated inhibition in transgenic tobacco has also been reported for TSWV, which is a negative, str&ed RNA virus (deHaan *et al.*, 1992). The smallest of the three genomic RNA of TSWV, the satRNA has an antisense strategy of gene expression. The genomic RNAs encode the 52 KDa nonstructural N gene whereas its complementary RNA encodes the viral nucleoprotein of 28.8 KDa. Expression of the N gene RNA transcript, which is of antisnse polarity verse the genomic RNA, resulted in RNA mediated resistance. Here translation of the N gene was not required for protection indicating that protection was due to RNA duplex formation. Pang *et al* (1993) demonstrated that several lines of plants transgenic for an untranslatable form of TSWV N gene showed a high degree of resistance against the homologous virus but no resistance to the distantly ones.

Table 2. Plant species & viruses for which RNA mediated resistance has been demonstrated.

Plant	Viruses	Reference
Potato	PLRV	Kawchuk *et al.* (1991)
	PVY	Smith *et al.* (1995)
Tobacco	TMV	Powell *et al.* (1989)
	TSWV, potyviruses	De Haan *et al.* (1992)
	TGMV	Day *et al.* (1991)
	TSWV	Pang *et al* .(1997), Prins *et al.* (1996 , 1997)
	TEV	Goodwin *et al.* (1996), Sneaney *et al.* (1996), Lindo *et al.* (1993)
	TYLCV	Bendahmane & Gronenborn (1997)
	BYMV	Hammond & Kamo (1995)
	PVX	Mueller *et al.* (1995)
	PVY	Smith *et al.* (1994), Mc Donald *et al.* (1997)
	CPMV	Sijen *et al.* (1996)
	TRSV	Yeps *et al* .(1996)

3. Mevement Protein Mediated Resistance

One mechanism of virus resistance in plants is the viability of viruses to move from cell to cell. Successful cell-to-cell transport of viruses depends on a combination of both host & virus encoded factors (Deom *et al*, 1992). For examples TMV encodes a 30 KDa movement protein (MP) that facilitates cell-to-cell movement by interaction with the interconnecting plansodesmata. When the TMV MP was expressed in transgenic tobacco, the macromolecular exclusion limit of the plasmodesmata was increased from approximately 800 Daltons to 9400 Daltons (Wolf *et al.*, 1989). Although there are many examples in which isolated protoplasts are susceptible to infection by a specific virus while whole plants of the same species & cultivars are resistant proteins which facilitate movement of virus particles from cell to cell or by long distance movement within a plant have been identified among the gene products of several plant viruses (Fujiwara *et al.*, 1993; Aitho *et al.*, 1993 & Meshi *et al.*, 1983). However in some cases there proteins have been localized at the plasmodesmata in plants transgenic for the movement protein (Atkin *et al.*, 1991 & Fujiwara *et al.*, 1993). When the MP was mutated to delete amino acids 3, 4 & 5 from the N-terminal in an infectious clone of TMV, the local & long distance movement of the virus was limited (Gafni *et al.*, 1992). Plasmodesmata of transgenic plants containing this dysfunctional movement protein were not modified to the same extent as these plants containing wild type movement protein. Also dysfunctional MP limited local & systemic movement of TMV & several other tobamoviruses. The effectiveness of movement protein mediated resistance (MPMR) was illustrated by the transgenic expression of viral movement proteins, which conferred resistance only when the transgene specified a dysfunctional MP (Lapidot *et al.*, 1993; Malyshenko *et al.*, 1993). Transgenic expression of functional MP either had no effect on virus infection or increased susceptibility. MP produced in transgenic plants could enable movement defective mutants of TMV to move to adjacent cells (Deon *et al.*, 1990. Holt & Beachy, 1991). It was suggested that that the transgenic expression of a dysfunctional MP would lead to competition for plasmodesmatal binding sites between the mutant MP & the wild type MP of the inoculated virus leading to the blockage of cell to cell movement of the virus (Lapidot *et al.*, 1993). An interesting & important aspect of MPMR is the broad-spectrum efficiency of the resistance mechanism (Deom *et al.*, 1992).

Table 3. Plant species & viruses for which MPMR mediated resistance has been demonstrated.

Plant	Viruses	Reference
Potato	PVX	Seppanen *et al* (1997)
	PAMV	
	PVM	
	PVS	
	PVX	Xu *et al* (1995)
	PLRV	Tacke *et al* (1996)
Tobacco	TMV	Malysheenko *et al*, 1993), Lapidot *et al* (1993), Cooper *et al* (1994, 1995)
	CPMV	Sijen *et al* (1995)
	TMoV	Duan *et al* (1997)

Plants containing this mutant protein of TMV also conferred varying levels of resistance to other members of virus group including caulimovirus, cucumber mosaic virus (Copper *et al.*, 1995). However, when functional movement was used the symptoms development was accelerated & severity of symptoms increased. Transgenic plants containing a mutant form of the 13 KDa gene of the triple gene was resistant to two potexviruses that were tested & potato virus S member of the Caulimovirus group, but not TMV (Beck *et al.*, 1994). The use of mutant MP to engineer virus resistance may be broad spectrum compared to other strategies. If this holds true it may be possible to engineer resistance to several viruses in a crop with a single construct.

In most plus sense RNA viruses, replication of viral RNA take place in the cytoplasm & this followed by transfer of newly synthesized genome from the site of replication to intercellular transport system. Interaction between movement proteins & replication proteins or nascent genomes may initiate this transport process & provide specificity for movement of viral RNA. This is followed by transport of viral genome to plasmodesmata & transmits through plasmodesmata. The MPs from complex with viral genome which can range in size.

B. PATHOGEN TARGATED RESISTANCE

Pathogen targeted resistance refers to molecular strategies that a viral function or component for inactivation. Elements used in these approaches may contain pathogen components. Such as gene sequences but the region active in pathogen disruption would be of viral origin.

1. Satellite RNAs

Satellite RNAs are entities associated with specific viruses that can replicate only in the presence of these (helper) viruses. Thus apparently do not code for any protein. Do not exhibit any homology to the genome of the helper virus, & are encapsidated by the helper virus coat protein. The helper virus on the other h& can replicate normally without the satRNA. Satellite RNA species have been found in association with a number of viruses. They are the RNA molecules like defective interfering (DI) RNA molecule & multiply in association with some viruses but are not strictly part of the virus genome. Recombination has been conserved between satellites RNA (Cascone *et al.*, 1990). After more underst&ing satellite RNA molecular biology, possibility to manipulate the sequences of these agents so that rose their transmissibility from the transgenic plant & yet retain their efficacy in ameliorating plant viral diseases. Thus it may lead to the use of satellite RNA sequences as virus resistance gene in a more attractive way. SatRNA can modify the expression of the symptoms produced by the helper virus in a host plant. In some cases, satRNA enhance the severity of symptoms in co injection with helper virus infection, & in other cases, the symptom are ameliorated (Palukaitis *et al.*, 1992). A probable risk of this strategy is that in transgenic plants these sat RNA could mutate during their infection, exhibit a shift from an alternating form to a virulent sat RNA. Moreover a satRNA or viruses producing a mild reaction on one host plant could elicit some symptoms on other host or in combination with a different starin of helper virus. However, practical experience with field trails in china has produced no evidence to support this hypothesis. In an eight genes study on tomato & pepper plants using mild or alternating combinations of CMV, satRNA in conjunction with CMV infections have not yet emerged (Tien & Wu, 1991).

In the past few years, satellite mediated virus control has been achieved by the development of transgenic plants expressing satellite sequences. Potential advantages over the use of pre-inoculaton include reduced labour requirements & the absence of a mild infection resulting from the pre-inoculating virus. The first example of engineered sat RNA mediated protection were reported in 1987 & the transgenic tobacco expressing CMV or TobRV sat RNA shown to be protected against the severe effects of their respective helper viruses (Harrison et al., 1987 & Gerlack et al., 1987). Since then, there have been several other reports of CMV sat RNA mediated resistance in transgenic tomato (Mcharvey et al., 1990; Tousch et al., 1990, Saito et al., 1992 & Mc Garvey et al., 1994).

The extent of specificity of satellite-mediated protection in transgenic plants varies depending on the satRNA strain & or host plant species. Tomato plants expressing S-CA RNA 5 CMV sat RNA sequences proved to be disease tolerant when infected by a number of CMV strains, but were not protected against the heterologous cucumoviruses, TAV (McGarvey et al., 1994). An additional advantage of multiple forms of protection has been reported by Yie et al. (1992), who found elevated resistance to CMV in transgenic tobacco expressing both a CMV coat protein gene & a satRNA of CMV that alternates virus symptom expression. This strategy appears to combine both the protective effects of CLMP & the interference of virus replication provided by the sat RNA. In field trails, Yie et al. (1992) found that chimeric plants provide twice the protection of plants individually transformed with either CMV coat protein or sat RNA. These finding suggest that it will be advantageous to utilize multiple sources of pathogen-derived resistance in plants when such sources are available. Since multiple artificial resistance genes may lead to both enhanced & more durable forms of resistance.

2. Plantibody Resistance

The expression of functional antibodies in plants was first reported in 1989 (Hiett et al., 1989) Animals unlike plants have evolved circulating immune systems which produce antibodies that can specifically recognize & bind pathogen molecules, targeting them for destruction. In case the light chain & heavy chain were expressed in separate plants & these plants were crossed to obtain plants that expressed both light & heavy chains. Many groups have since worked on expressing antibody genes in plants with very limited success. Van Engelen et al. (1994) have been able to produce functional antibodies in plants by cloning light & heavy chain genes in the same T-DNA but each derived by different promoter. In this way both genes were expressed in a coordinated manner in the same cell types. This resulted in assembly of functional antibodies in the plant. The antibodies accumulated in these plants were measured at 1.3% of the total leaf protein. Antibody expression in plants has also been obtained by the single expression of the immunoglobulin heavy chain variable domain (Benvennto et al., 1991). The domain alone is capable of biding antigen with good affinity, thus need for cloning & integrating both the light & heavy chain of an antibody can be avoided (Ben Vennto et al., 1991; Ward et al., 1992). Single chain antibodies (ScFv) consists of the variable domains of the heavy & light chains joined by a flexible peptide linker. These antibodies have the advantage of always being expressed in equimolar amounts at the same time & site avoiding problems with assembly. They are relatively small molecules (~30kDa) & can be cloned using the polymerase chain reaction making use of primers to conserved region of the IgG genes (Davis *et al.*, 1991). Recently, ScFv specific form the coat protein of artichoke mottle crinkle tombusvirus have been expressed in tomato plants (Taviadoraki *et al.*, 1993). The presence of the ScFv resulted in lower titers in systemically infected leaves & a delay in symptoms expression. These are some advantages to using ScFv for engineering resistance to plant viruses.

1) There are no viral sequences being expressed in the transgenic plant. Thus negates the concerns about the risk associated with expressing viral genes in plants.
2) The ScFv approach could be applied at any stage of the life cycle of the virus. At an early stage, for example, antibodies that interact with genes that are expressed in low copy number during virus infection (ie antibodies against the replicase for many viruses should be more effective than antibodies directed against the coat protein.

Antibodies that specifically bind replicase proteins might inactivate these proteins & prevent replication. The possible mechanism suggested that the interaction of the antibody with Ca 2+ binding sites could disrupt the uncoating of the virus or the assembly of the viral progeny.

3. Ribosome Inactivating Protiens (RIP)

Many proteins have been found to contain antiviral proteins, commonly termed as ribosome inactivating proteins. These proteins inhibit the translocation step of translation by catalytically removing a specific adenine base pair from 28 s ribosomal RNA. They re synthesize either pre or pre-pro proteins (Curzanige, 1994) & targeted to vacuoles because of their specific intracellular localization. RIPs do not affect endogenous 28 s RNA. It is supposed that RIPs enter cells together with the viruses & exert the damage to the host ribosome or possibly viral RNA (Broekaert, 2000). When purified RIPS are mixed with viruses & applied on plants, virus multiplication, & symptom development are dramatically suppressed. Use of several RIPs like pokeweed antiviral protein (PAP), trichosanthin, dianthin etc. have been explored recently for developing a broad-spectrum resistance in transgenic plants. For example potato plants transformed with PAP/PAP II genes exhibited resistance not only against the viruses like TMV, PVX, & PLRV but also against fungus (*Rizoctonia solani*) infection (Lodye *et al.*, 1993, Wang *et al.*, 1998).

In another strategy, a ribozyme was designed to trans splice the coding sequence of diphtheria toxin. A chain in frame with the CMV coat protein mRNA. The ribozyme was designed based on the self-slicing group I introns. Diphtheria toxin inhibits eukaryotic protein translocation. When yeast was used as model system, it was found that growth of these cells expressing the viral mRNA was specifically inhibited. Therefore, this strategy would be highly useful to develop resistance against viruses in plant with reduced illegitimate toxicity *in vivo* (Ayre *et al.*, 1999). By use of the PVX subgenomic RNA promoter & diphtheria toxin mRNA transgenic tobaccos showed a 20 fold reduction in PVX concentration (Wilson, 1993).

GroEL-mediated Resistance:

A homologue of GroEL, a chaparonin, produced by endosymbiotic bacteria from the whitefly vector *B. tabaci* was shown to bind with high affinity to the coat protein of *Tomato yellow leaf curl virus –Israel* (TYLCV-IL (Morin *et al.*, 1999). It was proposed that it may protect the virus from destruction during its passage through the insect's haemolymph (Morin *et al.*, 1999). This property was used as a tool to trap or from plant extracts for diagnostic purposes (Akad *et al.*, 2004). Given that the GroEL binds a wide range of begomovirus coat proteins it was really only a matter of time before the protein was put to the test as a resistance transgene.the GroEL transgene seemed to protect plants from the harmful effects of TYLCV-IL infections without substantially reducing virus titres within infected plants. Viral DNA leves were detectable in transgenic plants six weeks after virus challenge & were comparable to those in non –transgenic susceptible plants. Thia is a potential drawback to this mechanism in that resistant transgenic plants were apparently as good as infected non-mediated transmission (Akad *et al.*, 2007).

Recenlty Edelbaum *et al.* (2009) have shown that when they have expressed the GroEL gene in Nicotiana benthamiana plants, postulating that upon virus inoculation, GroEL will bind to virions, thereby interfering with pathogenesis. The transgenic plants were inoculated with the begomovirus tomato yellow leaf curl virus (TYLCV) & the cucumovirus cucumber mosaic virus (CMV), both of which interacted with GroEL in vitro, & with the trichovirus grapevine virus A (GVA) & the tobamovirus tobacco mosaic virus (TMV), which did not. While the transgenic plants inoculated with TYLCV & CMV presented a high level of tolerance, those inoculated with GVA & TMV were susceptible.

While the transgenic plants inoculated with TYLCV & CMV presented a high level of tolerance, those inoculated with GVA & TMV were susceptible. The amounts of virus in tolerant transgenic plants was lower by three orders of magnitude than those in non-transgenic plants; in comparison, the amounts of virus in susceptible transgenic plants were similar to those in non-transgenic plants. This study demonstrated that multiple resistances to viruses belonging to several different taxonomic genera could be achieved. Moreover, it might be hypothesized that plants expressing GroEL will be tolerant to those viruses that bind to GroEL in vitro, such as members of the genera Begomovirus, Cucumovirus, Ilarvirus, Luteovirus, & Tospovirus.

REFERENCES

[1]   Atkins, D., hull, R., Wells, B., Roberts, K., Moore, P. & Beachy, R.N. (1991). The tobacco plant is localized to plasmodesmata. *J Gen Virol* 72: 209-211.

[2]   Beachy, R.N., Loesch-Fries, S. & Tumer, N.E. (1990). Coat protein mediated resistance against virus infection. *Annu. Rev. Phytopathol* 28: 451-462.

[3]   Beck, D.L., Van Dolleweed, C.J., Longh, T.J., Balmori, E., Voot, D.M., &erson, M.T., O'Brein, I.E.W. & Forster, R.L.S.(1994). Disruption of virus movement confers broad spectrum resistance against systemic infection by plant virus with a triple gene block. *Proc Natl. Acad. Sci.* USA 91: 10310-10314.

[4]   Benveruto, E., Ordas, R.J., Tarazza, R., Ancora, G., Biocca, S., Cattaneo, A & Galeffi, P. (1991). Phytoantibodies: a general vector for the expression of immunoglobulin domains in transgenic plants. *Plant Mol. Biol.* 17: 865-870.

[5]   Braun, C.J. & Hemenway, C.L. (1992). Expression of amino terminal portions or ful length viral replicase genes in trabnsgenic plants confers resistance to potato virus X infection. *Plant Cell* 4: 735-738.

[6]   Chapman, S., Hills, G.T., Watts, J., & Baulcombe, D.C. (1992). Mutational analysis of the coat protein in potato virus X: effects of virion morphology & viral pathogenicity. *Virology* 191: 223-230.

[7]   Cooper, B., Lapidot, M., Heick, J.A., Dodds, J.A. & Beachy, R.N. (1995). A defective movement protein of TMV in transgenic plants confers resistance to multiple viruses whereas the functional analog increases susceptibility. *Virology* 206: 307-313.

[8]   Davis, G.T., Bedzyk, W.D., Voss, E.W. & Jacobs, T.W. (1991). Single chain antibody (SCA) encoding genes: one step construction & expression in eukaryotic cells. *Bio/Technology* (: 165-169.

[9]   Doem, C. M., Lapidot, M. & Beachy, R.N. (1992). Plant virus movement proteins. *Cell* 69: 221.

[10]   Fujiwara, T., Giesman-Cookmeyer, D., Ding, B., Lommel, S.A., & Lucas, W.J. (1993). Cell to cell trafeficking of macromolecules through plasmodesmata potentiated by the red clover necrotic mosaic virus movement protein. *Plant Cell* 5: 1783-1794.

[11]   Gafni, R., Lapidot, M., Berna, A., Holt, C.A., Deom, C.M. & Beachy, R.N. (1992). Effects of terminal deletion mutations on the function of the movement protein of tobacco mosaic virus. *Virology* 187: 499-507.

[12]   Gerlach, W.L., Llewellya, D. & Haseloff, J. (1987). Construction of a plant disease resistance gene from the satellite RNA of tobcco ringsopt virus . *Nature* 328: 802.

[13]   Haitt, A., Cafferkey, R. & Bowdish, K. (1989). Production of antibodies in transgenic plants. *Nature* 342: 76.

[14]   Harrison, B.D., Mayo, M.A. & Baulcombe, D.C. (1987). Virus resistance in transgenic plants that express cucumber mosaic virus satellite RNA. *Nature* 238: 799-802.

[15]   Kaniewski, W., Lawson, C., Sammons, B., Haley, L., Hart, J., Delannag, X. & Tumer, N.E. (1990). Filed resistance of transgenic Russet Burban pototo to effects of infection by potato virus Y. *Bio/Technology* 8: 750-753.

[16]   Kawchuk, L.M., Martin, R.R. & MvPherson, J. (1991). Sense & antisense RNA mediated resistance to potato leafroll virus in russet Burbank potato plants. *Mol. Plant-Microbe Interact.* 4: 247-253.

[17]   Lapidot, M., Gafny, R., Ding, B., Wolf, S., Lucas, W.J. & Beachy, R.N. (1993). A ysfunction movement protein of tobacco moaic virus that partially modifies the plasmodesmata & limits virus spread in transgenic plants. *Plant J.* 4: 959-963.

[18]   Lawson, C., Kaniewski, W.K., Haley, L., Rozman, R., Nowell, C., S&ers, P & Tumer, N.E. (1990). Engineering resisance to mixed infection in a commercial potato cultivar: resistance to potato virus X & potato virus Y in transgenic Russet Burbank. *Bio/Technology* 8: 127-132.

[19]   Malysheuko, S.I., Kondakora, O.A., Nazarova, J.V., Kaplan, I.B. Taliansky, M.E. & Atabekor, J.G. (1993). Reduction of tobacco mosaic virus accumulation in transgenic plants producing non functional viral transport proteins. *J Gen Virol* 74: 1149-1155.

[20]   McGarvey, P.B., Montasser, M.S. & Kaper, J.M. (1994). Transgenic tomato plants expressing satRNA are tolerant to some strains of cucumber mosaic virus. *J Am. Soc Hortic. Sci* 119: 642.

[21] Namba, S., Ling, K., Gonsavles, C., Slightom, J.L. & Gonsalves, D. (1992). Protection of transgenic plants expressing the coat protein gene of watermelon mosaic virus II or zucchini yellow mosaic virus against six potyviruses. *Phytopathlogy* 82: 940-945.

[22] Nelson, R.S., Powell-Abel, P & Beachy, R.N. (1987). Lesions & virus accumulation in inoculated transgenic tobacco plants expressing the coat protein gene of tobacco mosaic virus. *Virology* 158: 126-131.

[23] Pang, S.H., Nagpala, P., Wang, M., Slightom, J.L. & Gonsalves, D.(1992). Resistance to heterologous isolates of tomato spotted wilt virus in transgenic tobacco expressing its nucleocapsid protein gene. *Phytopathology* 82: 1223-1228.

[24] Quemada, H.D., Gonsalves, D. & Slightom, N. (1991). Expression of coat protein gene from cucumber mosaic virus srain C in tobacco: protection against infection by CMV strains trans,mitted mechanically of by aphids. *Phytopathology* 81: 794-799.

[25] Saito, Y., Komari, T., Masuta, C., Hayashi, Y., Kumashiro, T. & Takanami, Y. (1992). Cucumber mosaic virus tolerant transgenic tomato plants expressing a satellite RNA. *Theor. Appl. Genet.* 83: 679-685.

[26] Sit, T.L., Leclere, D. & Abouhaidar, M.G. (1994). The minimal 5' sequence for in vitro inititation of papaya mosaic potexvirus assembly. *Virology* 199: 238-242.

[27] Taviadoraki, P., Benvenuto, M., Trinca, S., De Martinis, D. Cattanco, A. & Galaffi, P. (1993). Transgenic plants expressing a functional single chain Fv antibody are specifivcaly protected from virus attech. *Nature* 366: 469-472.

[28] Tousche, D., Jacquemond, M. & Tepfer, M. (1990). Transgenic tomato plants expressing a cucumber mosaic virus satellite RNA gene: inoculation with CMV induces lethal necrosis. *C.R. Acad, Sci*, Paris, III, 311-377.

[29] Wolf, S., Deom, C.M., Beachy, R.N. & Lucas, N.L. (1989). Movement protein of tobacco mosaic virus modifies plasmodesmatal size exclusion limit. *Science* 246: 337.

[30] Yie, Y., Zhao, F., Zhao, G.Z., Liu, Y.Z., Liu, Y.L. & Tien, P. (1992). High resistance to cucumber mosaic virus conferred satellite RNA & coat protein in transgenic commercial tobacco cultivar G-140. *Mol. Plant microbe Interac.* 5: 460-463.

[31] MacKenzie, D.J. & Ellis, P.J. (1992). Resistance to tomato spotted wilt virus infection in transgenic tobacco expressing the viral nucleocapsid gene. *Mol. Plant-Microbe Intrec.* 5:285-291.

[32] Powell-Abel, P., Nelson, R.S., De, B., Noffman, N., Rogers, S.G., Fraley, R.T. & Beachy, R.N. (1986). Delay of disease development in transgenic plants that express the tobacco mosaic virus coat protein gene. *Science* 232: 738-743.

[33] Cuozzo, M., O'Connel, K.M., Kaniewski, R.X., Fang, N., Chau, H. & Tumer, N.E. (1988). Viral protection in transgenic tobacco plants expressing the cucumber mosaic virus coat protein or its antisense RNA. *Bio/Technology* 6:549-557.

[34] Day, A.G., Bejarano, E.R., Buck, K.W., Burrell, M. & Lichtenstein, C.P. (1991). Expression of an antisense viral gen in transgenic tobacco confers resistance to the DNA virus tomato golden mosaic virus. *Proc. Natl. Acad. Sci. USA* 88:6721-6725.

[35] Gielen, J.J.L., deHann, P., Kool, A.J., Peters, D., van Grinsven, M.Q.J.M. & Goldbach, R.W. (1991). Engineered resistance to tomato spotted wilt virus, a negative str& RNA virus. *Bio/Technology* 9:1363-1367.

[36] Hammond, J. & Kamo, K.K. (1993). Transgenic coat protein & antisense RNA resistnce bean yellow mosaic potyvirus. *Acta Hort.* 336:171-178.

[37] Harrison, B.D., Mayo, M.A. & Baulcombe, D.C. (1987). Virus resistance in transgenic plants that express cucumber mosic virus virus satellite RNA. *Nature* 328:799-802.

[38] Jongedjik, E., de Schutter, A.A.J.M., Stolte, T., van den Elzen, P.J.M. & Cornelissen, B.C.J. (1992). Increased resistance to potato virus X & preservation of cultivar properties in tragsneic potato under field conditions. *Bio/Technology* 10: 422-429.

[39] Citovsky, V. & Zambryski, P. 1991. How do plant virus nucleic acids move through intercellular connections? *BioEssay* 13:373-379.

[40] De Haan, P., Geilen, J.J.L., Prins, M., Wijkamp, I.G., van Schepen, A., Peters, D., van Grinsven, M.Q.J.M. & Gorlich, R. (1992). Characterization of RNA mediated resistance to tomato spotted wilt virus in transgenic tobacco plants. *Bio/Technology* 10:1133-1137.

[41] Palikaitis, P., Roossinck, M.J., Dietzen, R.G. & Francki, R.I.B. (1992). Cucumber mosaic virus. *Adv. Virus Res.* 41: 281-348.

[42] Sanford, J.C. & Johnston, S.A. (1985). The concept of parasite derived resistance deriving resistance genes from the parasite's own genome. *J. Theor. Biol.* 113:395-405.

[43] Tien, P. & Wu, G. (1991). Satellite RNA for the biocontrol of plant disease. *Adv. Virus Res.* 39: 321-431.

[44] Nelson, R.S., Powell-Abel, P. & Beachy, R.N. (1987). Lesions & virus accumulation in inoculated transgenic obacco plants expressing the coat protein ene of tobacco mosaic virus. *Virology* 158:247-251.

[45] Loesch-Fries, L.S., Merlo, D., Zinnen, T., Burhop, L., Hill, K., Krahn, K., Jarvis, N., Nelson, S. & Halk, E. (1987). Expression of alfalfa mosaic virus RNA 4 in transgenic plants confers virus resistance. *EMBO J.* 6: 1845-1851.

[46] Hemenway, C., Fang, R., Kaneiwski, W.K., Chau, N.H. & Tumer, N.E. (1988). Analysis of the mechanism of protection in transgenic plants expressing the potato virus X coat protein or its antisense RNA. *EMBO J.* 7: 1273-1289.

[47] Lawson, C., Kaniewski, W., Haley, L., ROzman, R., Newell, C., S&ers, P. & Tumer, N.E. (1990). Engineering resistance to mixed virus infection in commercial potato cultivars: resisatance to potato virus X & potato Y in transgenic Russet Burbank. *Bio/Technology* 8: 127-134.

[48] Namba, S.K., Ling, C., Gonsalvez, C., Sligthom, J.L. & Gonsalves, D. (1992). Protection of transgenic plants expressing the coat protein of watermelon mosaic virus II or zucchini yellow mosaic virus against six potyviruses. *Phytopathology* 82: 940-945.

[49] Ploeg, A.T., Mathis, A., Bol, J.F., Brown, D.J.F. & Robinson, D.J. (1993). Susceptibility of transgenic tobacco plants expressing tobacco rattle virus coat protein to nematode transmitted & mechanically inoculated tobacco rattle virus. *J. Gen. Virol.* 74:2709-2715.

[50] Linbdo, J.A., Silva-Rosales, L., Proebsting, W.M. & Dougherty, W.G. (1993). Induction of highly specific antiviral state in transgenic plants: Implications for regulation of gene expression & virus resistance. *Plant Cell* 5: 1749-1759.

[51] Wolf, S., Doem, C.M., Beachy, R.N. & Lucas, W.J. (1989). Movement protein of tobacco mosaic virus modified plasmodesmatal size exclusion limit. *Science* 246: 377-379.

[52] Akad, F., Eybishtz, J.M., Edelbaum, D., Gorovits, R., Dar-Issa, O., Iraki, N. & Czosnek, H. (2007). Making a friend from a foe: expressing a GroEL gene from the

whitefly Bemisia tabaci in the phloem of tomato plants confers resistance of tomato yellow leaf curl virus. *Arch Virol.* 152: 1323-1339.

[53] Morin, S., Ghanim, M., Zeidan, M., Czosnek, H., Verbeek, M., ban den Heuvel, J.F.J.M. A GroEL homologue from endosymbiotic bacteria of the whitefly *Bemisia tabaci* is implicated in the circulative transmission of tomato yellow leaf curl virus. *Virology* 256:75-84.

[54] Akad, F., Dotan, N. &Czosnek, H. (2004). Trapping of tomato yellow leaf curl virus (TYLCV) & other plant viruses with a GroEL homologue from the whitelfy *Bemisia tabaci. Arch. Virol.* 149: 1481-1479.

[55] Edelbaum, D., Gorovits, R., Sasaki, S., Ikegami, M. & Czosnek, H. (2009). Expressing a whitefly GroEL protein in *Nicotiana benthamiana* plants confers tolerance to tomato yellow leaf curl virus & cucumber mosaic virus, but not to grapevine virus A or tobacco mosaic virus. *Arch Virol.* 154: 399-407.

In: Recent Trends in Biotechnology and Microbiology
Editors: Rajarshi Kumar Gaur et al., pp. 43-53

ISBN: 978-1-60876-666-6
2010 Nova Science Publishers, Inc.

Chapter 4

MICROARRAYS: CONCEPT AND APPLICATIONS

Smita Singh[1], Pawas Goswami[1], Kanti P. Sharma[2] and R. K. Gaur[2]

Department of Microbiology, MDS University, (Ajmer), Rajasthan, India[1]
Department of Science, Faculty of Arts Science and Commerce,
Mody Institute of Technology and Science,
Luxmangarh, Sikar (Rajasthan),India[2]

ABSTRACT

Microarrays have emerged as a powerful gene expression analysis technique that deals with altered phenotypes at gene expression level. Its ability to correlate various molecular events with gene expression diversify its applications to the fields of pharmaceutical industry, clinical laboratories, in drug development, toxicity assessment, and other applications based on the comparison of expression patterns of genes specific to a particular tissue or genes involved in a metabolic pathway.

INTRODUCTION

With the completion of first draft of human genome project in the year 2001, huge amount of valuable information regarding entire human genomic DNA sequence has added new dimensions to the field of molecular biology and biotechnology research with the possibility of addressing significant novel issues like new genes identification, targeting their function, correlating genes with the disease phenotype and elucidating the correlation between genotype and phenotype (Kurella, 2001). Soon afterwards it has become feasible to address these issues quickly, due to invention of a new technological advancement, in the form of Microarrays, which is simply a robust *ex situ* "Northern hybridization" assay, taking place on miniature sized gridded solid support as platform to examine the expression pattern of a large no of genes at a time. These are ceramic or silicon wafers or very thin glass slides, with immobilized arrays of short DNA probes in the form of regular pattern of spots on their surface. The main advantage of microarray chips is that these chips replace cumbersome equipments with miniatured, microfluidic assay chemistries and provides ultra sensitive qualitative as well as quantitative detection of gene expression at significant lower cost per assay than traditional methods, in a significantly much smaller amount of space. Microarrays can be highly beneficial in solving several puzzling queries in the field of biomedical, pharmaceutical and clinical research, such as: exposing important disease mechanisms, identifying novel drug targets and predicting therapeutic response to drugs. These entire phenomenons are basically linked to individual genetic makeup and changes occurring at gene expression level (Isakkson and Landergran, 1999). Microarrays can also be used to characterize the functions of novel genes, identify genes in a biological pathway and for the analysis of genetic variation. Moreover, the expression profile of the genome can be used as a disease "fingerprint." The strength of this technique lies in its capacity to monitor even whole genome on a single chip, thus getting a better insight into the coordinated expression of thousands of genes simultaneously..However, when visualized in technical terms the merits of this technique are as follows-

1) The use of solid non-porous surface is particularly advantageous, as it

    - Enables the deposition of very small amounts of biochemical material at a precisely defined position and checks the diffusion of applied material into the supporting matrix.
    - Prevents the absorption of reagents and sample into the matrix, thus allowing the rapid removal of organic and fluorescent compound during biochip fabrication and use.
    - Permits the use of small sample volumes enabling high sample concentrations and rapid hybridization kinetics that increases the quality of the array elements.
    - Glass substrates permit the use of much smaller reaction volumes (5-200ul) than are used in the traditional filter assays (5-50ml).
    - Glass substrates also allow the use of cover slips sealed in reaction chambers for the hybridization reaction.

2) Small reaction volume

    - Reduces reagent consumption and cost.
    - Increases the concentration of nucleic acid reactants in micro array assays (0.1-1mM) by as much as 100,000 fold compared to conventional assays.
    - Accelerates hybridization kinetics and thus reduces the length of time needed to obtain a strong fluorescent signal.

TYPES OF MICROARRAYS

On the basis of the required application, microarrays are designed either as partial or complete cDNA arrays, DNA or oligonucleotide microarrays.

Partial or Complete cDNA Arrays

It comprises of DNA copies of transcribed genes, only ranging from 500-3000 nucleotides in length. It is used for relative quantitative measurement of gene expression of two or more sources. These arrays are more specific then shorter oligonucleotide probes and 2 or 3 spots cover sequence of an entire gene.

Oligonucleotides or DNA Microarrays

These are the short synthesized fragments of DNA, usually 25 to 60 mer in size, and are used for identification of altered gene forms i.e. to provide high degree of discrimination of single mismatch. Due to their short length, number of spots representing one gene is greater in this pattern. These densely packed arrays are particularly useful in the determination of

- Gene alleles.
- Single nucleotide polymorphism.
- Other polymorphisms in an individual's genome.

These are more expensive, as DNA sequences are directly synthesized on to the chip.

Steps Involved in Setting up of a Microarray Experiment
The establishment and completion of a microarray experiment is carried out in following steps-

1) Fabrication of microarray.
2) Sample preparation.
3) Hybridization with probe and Data analysis.

1) **Fabrication of Microarray:** It entails of

    a) Probe DNA preparation
    b) Immobilization of probe DNA on to the chip.

Probe DNA Preparation
Once a desired gene sequence data is obtained from public database/repositories or from institutional sources, it is cloned individually by using universal primers in tens of thousands of mass PCR reactions, performed by robotic systems, to produce millions of identical copies through amplification. Probes are either-

- Genomic DNA/
- Expressed sequence tags/
- Synthetic oligonucleotide (Lipshutz, *et al.*, 1999).

Immobilization of Probe to the Chip

DNA is spotted, printed or directly synthesized on to the support. For this purpose a number of techniques are being used now a days by different commercial companies.

On Chip Synthesis/Photolithography

This is a highly advanced technology based process that makes use of U.V. light to covalently attach the DNA strands to the slide (Fodor, et al., 1991). In this, nucleotides are synthesized one by one directly on to the chip on desired location, so that both the sequence and size of the probe entirely remain in control. Besides other advantage associated with this method is that no PCR steps are required and areas of specificity can be chosen separately within any given cDNA or given genomic sequence. The synthesis of microchip starts with binding of photosensitive modified deoxy nucleotide precursors protected at their 3'OH end by linkage with phosphodegenerated acid (PGA) to the surface of microchip in discrete array locations. Function of PGA is to make DNA chemically inert. Now U.V. light slightly below 400 nm range is passed through expensive chrom or glass masks on only chosen locations to achieve photo desensitization of PGA and hence removal from these sites. During this step the 3' hydroxyl end becomes free and ready to bind with the next nucleotide. Selected nucleotide is then flushed over the chip and will bind only to the locations that have had the PGA removed. The chip is flushed with PGA again so as to make the entire chip chemically inert once again. This process continues with a new mask each time in order to create a library of DNA strands on a single chip. However, NimbleGen has devised Maskless Array Synthesis (MAS) technology, which uses digital light processing and rapid, high-yield photochemistry to synthesize DNA and it enjoys the exclusive worldwide license to the MAS technology from the Wisconsin Alumni Research Foundation. The major drawback of this method is that the oligo size obtained is very short, only up to 25 mer, so hybridization is less efficient and this method requires more precise knowledge of the sequence.

Mechanical Spotting

In mechanical spotting, a robotic arrayer, picks up nanoliter quantities of DNA probes and deposits them with the aid of capillary printing tips either in the format of split quills or solid rods by direct contact on to the slide surface. To avoid cross contamination, washing is done after every single use between applications. By using robotic arrayer and capillary printing tips, at least 23,000 gene fragments can be printed on to a microscopic slide.

Ink Jet

Ink Jet printing technology was adapted from the personal computer printer industry. A DNA sample is loaded into a tiny nozzle equipped with a piezoelectric device. The device expels a precise volume of DNA sample from the nozzle on the slide surface. The nozzle is washed after each deposition to provide a clean nozzle ready for next sample. After deposition on the solid matrix, spots are immobilized by exposing to the UV radiation, which cross-links the probe DNA fragments, and fix them on the slide, and then slides are dipped into water heated at 95°C or treated with alkali for their denaturation. Glass support based arrays have the advantage that two-color fluorescence labeling is possible, with no inherent background fluorescence. The shortcoming of poor adherence to the glass surface is overcome by pretreatment of glass slide with poly lysine to enhance the adherence of DNA.

Ink Jet Style in Situ Synthesis

This method was developed by Rosetta inpharmatics and Agilent Technologies Inc. In this method, standard di methoxy trityl blocked phosphoramidites are used to construct oligonucleotides. The advantage with this method is that since reverse amidites i.e. 3'dimethoxy trityl blocked 5'phosphoramidites rather than 5' dimethoxy trityl blocked 5'phosphoramidites can be used to make oligonucleotides with free 3'OH, it allows the synthesis to proceed in 5'-3'direction like enzymatic synthesis. Therefore this kind of chip can be used for direct hybridization assays or for assays using extension of primer by polymerase enzymes after hybridization to the target DNA (Arrayed Primer Extension Assays). Furthermore coupling efficiency is higher and the quantity of oligonucleotides up to 60 mer in length is possible using the system.

2) Sample preparation

In the next step, mRNA from sample is extracted and copied to complementary DNA and later labeled with different fluorochromes by any of the following methods-

- Nick translation for genomic DNA.
- Labeled nucleotides in RT-PCR reaction to copy mRNA to cDNA.
- Random priming in PCR amplification.

3) Hybridization and Data Analysis

A robotic arrayer now supports the labeled target samples from two or different sources from microtiter plate wells to the microarray chip, using comb of needles or pins at precise locations. Complimentary annealing between the matched target and probe DNA on the microarray slide locks the target DNA on that precise location. Once the hybridization is complete, slide is removed from the hybridization chamber, and excess unbound DNA is washed. Now degree of hybridization and hence the level of expression is assessed by monitoring fluorescent emission using laser excitation from a scanning fluorescence microscope, focused between the surface of glass slides and the solution. As soon as each spot (an area of around 100 μ^2m) on the microarray slide is illuminated, the laser excites the fluorescence tag attached with the sample DNA and fluorescence intensity is measured separately for each different fluorochrome. Once the hybridization signal is measured, microscope and charge coupled device (CCD) camera work together to create a digital 2D image of the array with each spot being identified by it coordinates on the microarray. The data is now stored in the computer and a special software program is used to measure the signals as absolute intensities for a given target or as ratios of two probes with two different fluorescent labels representing two separate treatments to be compared or with one probe as an internal control. The ratio of two signals provides relative response ratios rather than an estimate of an absolute signal.

Applications

The ability of microarrays to probe differences in gene expression level under varying conditions makes microarrays the method of choice in many fields, such as disease fingerprinting, drug targeting and evaluation, toxicity assessment, signal transduction research and cancer treatment. The important applied fields for microarray may be short listed as-

Gene Expression Analysis

In the basic scheme of such kind of analysis, the array used is expression chip having arrayed probes of immobilized cDNA copies of known genes and is able to analyze the expression of both control and sample DNA from healthy and diseased tissues respectively by labeling two different source specific DNAs with different colored fluorescent dyes, usually Cy5 i.e. red dye for diseased tissue (sample) and Cy3 i.e. green for normal tissue DNA. Then both of these probes are mixed in equal amounts and hybridized on microarray plate. After hybridization, florescence signal is color coded, taking advantage of the fact that each type of fluorochrome attached, fluoresce at different wavelengths. Once the expression pattern of the genes involved in a known disease is identified, it can be compared with the expression profile of unknown diseased individual (Schena, et al., 1995). If both match, appropriate treatment for that disease can be initiated within no time. Further expression chips can also be very useful for the treatment of genetic diseases, such as cancer. The current cancer diagnosis is based on the examination of histological aspects, but this can not reveal the underlying genetic aberrations or biological processes that contribute to its malignancy. On the other hand DNA microarray diagnosis is itself a functional genomics aproach, which has enabled examination of expression profile of entire cancer cell population i.e. whole transcriptome simultaneously. The resulting altered expression profile can then be viewed as a blue print, by which that gene can be identified which affects the cellular function. As in the case of a second tumour in a patient (for example, lung cancer after a larynx cancer), expression patterns establish whether it is a metastasis of the first tumour (similar patterns will be produced) or a second primary tumour (different patterns will be produced). Similarly, in the case of a metastasis in a patient with two (or more) primary tumours, the origin of the metastasis can be traced, which is important for future treatment (Snijders, et al., 2000). In addition, expression patterns could be useful for the classification of tumours (Perou et al., 2000; Sorlie et al., 2001) and for the risk assessment of premalignant lesions. One of the major outcomes of microarray technique is that, through a specifically designed microarray experiment, not only the gene undergoing mutation can be targeted, but the particular time period at which this change occurs in the cell cycle, can also be estimated as well.

Principally the cell cycle data must be generated in multiple arrays first and referred to time zero. Analysis of the collected data would help in elucidating the details of "cell cycle" further and also its clock, So that the exact time and genes undergoing changes in expression can be targeted. This technique has been used in diagnosis of several cancers such including breast cancer, B cell lymphomas, leukemia, colon denocarcinoma, ovarian tumors (Ismail , et al., 2000; Mok, et al.,2001; Bayani, et al., 2002 ;Sridhar et al.,2001, 2002) and lung carcinoma (Altman et al., 2001;Garber et al. 2001;Meyerson et al., 1999).

Genomic Gains and Losses

It has been hypothesized that certain chromosomal gains and losses are related to cancer progression and that the pattern of these changes is relevant to clinical prognosis. One of the microarray scheme, comparative gene hybridization microarray (CGH) screens directly for genome wide detection of chromosomal gains and losses or for a change in the number of copies of a particular gene involved in the disease state. In CGH microarray, large piece of genomic DNA serve as the probe DNA and each spot has known chromosomal location. Rest of the experiment scheme is same as for basic gene expression experiment.

Gene Mutation/Genotyping

A more recent application for DNA chips occurs in the study of natural DNA variations among individuals called SNPs (Single Nucleotide Polymorphisms). Thousands of these SNPs can be examined in an individual to develop a type of genetic fingerprint with the help of microarray chips, only consisting of different range of SNPs for mutated genes (Saplosky et al., 1998). The SNPs within the heat shock protein Hsp60 gene have been used as species-specific markers to differentiate three *Campylobacter* species for the development of microarray based *Campylobacter* detection method (Zhang et al., 2000) This approach has a significant shortcoming also, that a separate microarray or chip is required to genotype every single individual.

Medical and Clinical Diagnosis

Disease diagnosis, prognosis and treatment are the exciting new areas to be explored using microarray technology in the near future. Particularly thrilling is the potential of DNA array technology to provide "individual" or "personalized" treatment for a wide range of clinical conditions, based upon unique fingerprint for each of these conditions. This approach is particularly important for many clinical applications especially those, that require mass testing, quick diagnosis and general diagnostics. For general blood screening this technique generates the possibility of quick comprehensive donor and recipient blood testing on a single chip for number of individuals, instead of number of one to one assays used traditionally. In addition, it can find application in testing many medical hypotheses concerning diverse human health issues related to alcoholism and drug abuse.

Pharmacogenomics

Pharmacogenomics is the concept of hybridization of functional genomics and molecular pharmacology. It aims to find correlation between therapeutic responses to drug and genetic profiles of patients. Thus pharmacogenomics is actually the study of entire complement of pharmacologically relevant genes, how they manifest their variations, how these variations interact to produce disease phenotypes and how these phenotypes affect drug response? Pharmacogenomics should reveal how an individual's genetic inheritance affects the body's response to drugs and holds the promise that drugs might be one-day tailor made. A person's response to medicines can vary with several factors like diet, age, life style and state of health, but understanding an individual's genetic makeup is thought to be the key to design personalized drugs with great efficacy and safety. There are three major targets, which are regular goals to be fulfilled by the pharmaceutical industry.

1) To discover a drug for an already defined target
2) To assess drug toxicity
3) To monitor drug safety and effectiveness

The most attracting feature of microarray technology for the benefit of pharmacological science and pharmaceutical industry is that it aims to target the effect of a potential drug candidate directly on the mass of simultaneous genes or clusters of genes involved in the disease progression, but not on the disease phenotypes, which is sometimes a very late phenomenon.

Hence treatment becomes more precise, early and suitable to the correct stage of genetic alteration, depending on the degree of advancement of the disease for a particular affected individual. Besides, expression chips also can be used to develop new drugs. For instance if a certain gene or gene cluster is over expressed in a particular disease, researchers can use expression chips to observe, if a new drug will reduce over expression and force the disease back into remission. Drug expression profiling could be used to classify potential therapeutic agents based upon the molecular mechanism of action.

Toxicogenomics

The goal of toxicogenomics is to find a correlation between toxic response to toxicants and changes in the genetic profile of the object exposed to such toxicants. Predictive toxicology is a field which can be highly benefited by microarray, (Afshari, et al., 1999). It can provide significant information related to drug toxicity, safety and effectiveness by determining the potential of a new compound early and thus can save both precious time and money by focusing precisely on those compounds, which are more likely to succeed.Thus the ability to predict the toxic effects of potential new drugs are crucial to prioritize only right compound as potential drug candidate and also eliminates costly failures in drug development. Thus microarray offers a boosting technology for the pharmaceutical industry to reap the benefits of a well targeted streamlined advanced research focus only on the potential drug candidates thus avoiding the wastage of the enormous amounts of money, time and human resources on the wrong compounds at an early stage.

Electronic Chip Technology: A Promising Advanced Version

An interesting development in recent years in microarray technology has emerged as nanochip. The nanochip exploits the charged nature of biological molecules. Each desired site on the nanochip is charged with platinum wires, as a result DNA and RNA molecules rapidly move and concentrate on these charged sites. Uncharged sites do not attract DNA or RNA. The main attracting feature of this technique is that the high concentrations of nucleic acids with electronics, enables rapid hybridization in mere two minutes instead of 16-18 hrs, in passive hybridization. Electronic hybridization and stringency is carried out with single base resolution (Zhang et al., 2006).

Drawbacks of Microarray Technology

In nearly two decades, since their conception DNA chips and microarrays have revolutionised genomic research in a big way and have forever changed the way in which molecular biologists approach their research. Although in principle it looks a very simple and sensitive technique but practically it has its own limitations. Firstly since it is an expression based technology, capable of monitoring only cellular responses at RNA level, it fails to identify the signaling changes occurring at protein or post translation level.

Secondly at technical front, microarray fabrication is also prone to human error. The sheer amount of information expressed on a microarray chip also opens the possibility of incorrect labeling. Simultaneously, if a mistake is made at the cloning or PCR amplification stage in the printing process, the error may be transmitted through the hybridization step and gene may be incorrectly identified. Before transferring clones from the microtiter plate, there is also the possibility of cross contamination into other wells during the preparation, leading to wrong DNA or mix of clones sometimes. At hybridization step care must be taken to reduce background noise (fluorescent signals from artifacts) as much as possible, in order to get correct quantitative measurement of the expression profile of the desired gene or subset of genes.

Errors in sequence databases sometimes lead to errors in gene annotations therefore; confirmation of clone identification is essential. For cDNA microarray technology routine sequencing may be advantageous in confirming the assumed identity of a gene of interest. Furthermore microarrays indicate quantitative assessment of the level of gene expression, a value that for some genes may be imprecise. Although qualitative interpretation may suffice for some investigation, however if precise quantitation is important as for clinical diagnostic applications the level of gene expression should be validated by another technology such as real time quantitative PCR.

Northern blots and ribonuclease protection assays also provide the benefit of determining not only a quantitative measure, but also the number of potential transcripts detected with the chip sequence. Another problem related with the microarray analysis is that if a cell culture or tissue is heterogeneous, significant gene changes may be related to only a small fraction of the cells. In this condition it is required to confirm localization with an *in-situ* technique. Furthermore one of the major issue to be solved with microarray assay is that microarray requires a standard of gene expression profile, to which abnormal cell profile can be compared. However, because all healthy individual differ in their gene expression profile, there is no standard for normal. Hence, there is no absolute control to which profiles can be compared. Many times, it is quite difficult to include the proper controls to ensure the changes in gene expression, truly due to the factor being investigated.

Commercial Companies in Microarray Research

Two of the major commercial companies, popular in microarray field are Affymatrix (Santa Clara, California) and Agilent (Palo Alto, California). Affymatrix provides 25 mer oligonucleotide arrays synthesized *in-situ* using solid phase chemical synthesis in combination with photolithography. It supplies commercial microarrays with symbol Gene Chip and with commercial name U133 for human genome microarrays containing more than one million different oligos representing over 33,000 best characterized human genes. Agilent on the other hand offers human, mouse and rat cDNA arrays and custom oligonucleotide arrays (25 or 60 mer) on 1x3" glass slide in 8,400 or 22,000 formats. Beside these two, some other companies also well known in this field are-

- Incite genomics (Palo Alto, California)
- Amersham Biosciences (Piscataway, New Jersey, USA.)
- Motorola (North Brook, Illinois, USA)
- Clontech Laboratories, Incorporation (Palo Alto, California, USA)
- Mergen limited (San Leandro, California, USA)
- Stratagene (La Jolla, California, USA)
- NimbleGen Systems, Incorporation (Madison, Washington, USA)

- Promega Corporation (Madison, Washington, USA)

Future Prospects

Microarray technology is being acknowledged as a corner stone in the development and integration of genetics, pharmaceuticals, clinical diagnostics and drug development on a common platform. The strength of the data generated, in relation to the genes involved, put these technology quantum leaps ahead of general approaches in these areas.

Although very high cost involved in the manufacture of microarrays put a serious limitation to its wide use, in near future; due to many recent advancements in technology related to microarray fabrication, data handling and processing tools have reduced its cost upto a certain level so that it can be afforded in at least some of well equipped, well funded laboratories world wide. The whole plethora of improved technologies, especially in bioinformatics will enhance its scope enormously and will pave its way up to common molecular biology laboratories soon.

As a result of its implementation, a personal series of microarray tests may find its place in the pathological labs to understand a disease at the individual level and to pick drugs tailored to the patients microarray profiles and hence highly specific for every new patient. The strength will depend upon quick results using minute quantities of diseased cells.

Thus although currently the microarray analysis of gene expression is in early stages, it is far far ahead than the traditional gene by gene methods, in solving the mystery of complex gene regulating network consisting of tens of thousands of genes. It is expected that, coupled with advanced informatics, microarray diagnostics will permit individualized investigation of the physiology and disease patterns and may be able to dispense the need of large, costly and often difficult to interpret clinical trials.

As a concluding note, it can be understood that microarray technology is improving to the point that soon, an entire genome will be monitored with very high resolution on a single chip. With such potentials, microarrays are certainly on their way to revolutionarize bioscience and to allow a deep insight directly into the gene. This will help to find the answers of all the complexities associated with alteration in physiological profile of any individual. In essence, it certainly beholds the promise to reach up to roots of the molecular genomic science, to create a landmark to be cherished for centuries.

REFERENCES

[1]   Afshari, C.A., Nuwaysir, E.F., Barrett, J.C. (1999). Application of Complementary DNA Microarray Technology to Carcinogen Identification, Toxicology and Drug Safety Evaluation. *Cancer Res*.59: 4759-4760.

[2]   Altman, R.B., Brown, P.O., Botstein, D., Petersen, I., Pacyna-Gengelbach, M., Van de Rijn, M., et al. (2001). Diversity of gene expression in adenocarcinoma of the lung. *Proc Natl Acad Sci*.98: 13784-13789.

[3]   Bayani, J., Brenton, J.D., Macgregor, P.F., Beheshti, B., Albert, M., Nallainathan, D, et al. (2002). Parallel analysis of sporadic primary ovarian carcinomas by spectral

karyotyping, comparative genomic hybridization and expression microarrays. *Cancer Res.*62: 3466-3476.

[4]   Fodor, S.P., Read, J.L., Pirrung, M.C., Stryer, L., Lu A.T., Solas, D. (1991). Light-directed, spatially addressable parallel chemical synthesis. *Science.*251: 767-773.

[5]   Garber, M.E., Troyanskaya, O.G., Schluens, K., Petersen, S., Thaesler, Z., Pacyna-Gengelbach, M., et al. (2001). Diversity of gene expression in adenocarcinoma of the lung. *Proc Natl Acad Sci U S A.* 98:13784-13789.

[6]   Isaksson, A., Landergran, U. (1999). Accessing genome information: alternatives to PCR. *Curr Opin Biotechnol.* 10(1): 11-15.

[7]   Ismail, R.S., Baldwin, R.L, Fang, J., Browning, D., Karlan, B.Y., Gasson, J.C., et al. (2000). Differential gene expression between normal and tumor-derived ovarian epithelial cells. *Cancer Res.* 60: 6744-6749.

[8]   Kurella, M., Hslao, Li., Yoshida, T., Randall, J.D., Chow, G., Sarang, S.S., Jensen, R.V., Gullans, S.R. (2001). DNA microarray analysis of complex biologic processes. *J. Am Soc Nephrol.* 12(5): 1072-78.

[9]   Lipshutz, R.J., Fodor, S.P, Gingeras, T.R., Lockhart, D.J. (1999). High density synthetic oligonucleotide arrays. *Nat Genet.* 21: 20-24.

[10]   Meyerson, M., Mark, J., Lander, E.S., Wong, W., Johnson, B.E., Golub, T.R., Sugarbaker, D.J., Ladd, E.C., Beheshti, J., Bueno, R., Gillette, M., Loda, M., Weber, G., Bhattacharjee, A., Richards, W.G., Staunton, J., Li, C., Monti, S., Vasa, P.(1999). Classification of human lung carcinomas by mRNA expression profiling reveals distinct adenocarcinoma subclasses. *Cancer Res.* 59: 4759-4760.

[11]   Mok, S.C., Chao, J., Skates, S., Wong, K., Yiu, G.K., Muto, M.G., et al. (2001). Prostasin, a potential serum marker for ovarian cancer: identification through microarray technology. *J Natl Cancer Inst .*93:1458-1464.

[12]   Perou, C.M., Sorlie, T., Eisen, M.B., van de Rijn, M., Jeffrey, S.S, Rees, C.A, et al. (2000). Molecular portraits of human breast tumours. *Nature.* 406:747-752.

[13]   Sapolsky, R., Hsie, L., Berno, A., et al. (1998). High-throughput polymorphism screening and genotyping with high-density oligonucleotide arrays. *Genetic Analysis: Biomol Engg.* 14: 187–192.

[14]   Schena, M., Shalon, D., Davis, R.W., Brown, P.O. (1995). Quantitative monitoring of gene expression patterns with a complementary DNA microarray. *Science.* 270: 467-470.

[15]   Shridhar, V., Lee, J., Pandita, A., Iturria, S., Avula, R., Staub, J., et al. (2001). Genetic analysis of early- versus late-stage ovarian tumors. *Cancer Res.* 61: 5895-5904.

[16]   Shridhar, V., Sen, A., Chien, J., Staub, J., Avula, R., Kovats, S., et al. (2002). Identification of underexpressed genes in early- and late-stage primary ovarian tumors by suppression subtraction hybridization. *Cancer Res.* 62: 262-270.

[17]   Snijders, A.M., Meijer, G.A., Brakenhoff, R.H., van den Brule, A.J.C., van Diest, P.J. (2000). Microarray techniques in pathology: tool or toy? *Mol Pathol.* 53 (6): 289–294.

[18]   Sorlie, T., Perou, C.M., Tibshirani, R., Aas, T., Geisler, S., Johnsen, H., et al. (2001). Gene expression patterns of breast carcinomas distinguish tumor subclasses with clinical implications. *Proc Natl Acad Sci U S A.* 98:10869-10874.

[19]   Zhang, H., Zhilong, G., Pui, O., Liu, Y., Li, X.F. (2006). An electronic DNA microarray technique for detection and differentiation of viable *Campylobacter* species. *Analyst.* 131: 907-915.

In: Recent Trends in Biotechnology and Microbiology
Editors: Rajarshi Kumar Gaur et al., pp. 55-63

ISBN: 978-1-60876-666-6
2010 Nova Science Publishers, Inc.

Chapter 5

BIOLOGICAL CHARACTERIZATION, GENOME ORGANIZATION AND MANAGEMENT STRATEGIES OF BEGOMOVIRUSES CAUSING SEVERE ECONOMIC LOSSES TO CHILLI CULTIVATION IN INDIA

M. S. Khan, S. K. Snehi and S. K. Raj[*]

Molecular Plant Virology, Centre for Plant Molecular Biology (CPMB) division,
National Botanical Research Institute, Rana Pratap Marg, Lucknow-226 001, U. P. , India

ABSTRACT

Chilli pepper (*Capsicum annuum*), one of the most important spice and vegetable crops of the world, is grown in an area of 880,000 hectares in India with a production of 1.2 million tonnes (Anonymous, 2005). India is the world's largest producer, consumer and exporter of chillies to a number of countries viz. Sri Lanka, Bangladesh, South Korea, USA, Germany, Japan, UK and France. In India, Andhra Pradesh, Orissa, Maharashtra, West Bengal, Karnataka, Rajasthan, Tamil Nadu and Uttar Pradesh are the major chilli growing states. The crop is grown throughout the year and in some places overlapping crops are also taken up. Such a cropping system also provides ideal conditions for the perpetuation and spread of viruses and their vectors. It is, therefore, natural that plant viruses have emerged as a major constraint in improving the production of chilli.

INTRODUCTION

Viruses belonging to Cucumo-, Gemini-, Nepo-, Poty-, Tobamo- and Closterovirus group are known to infect chilli under natural conditions in various part of the world/country. Most damaging of these are the whitefly (*Bemisia tabaci*) transmitted geminiviruses (WTGs), which form a major genus begomovirus of the family *Geminiviridae*. The begomovirus

---

[*] Email: skraj2@rediffmail.com

causing leaf curl disease of chilli are a major limiting factor and threat to chilli cultivation in the tropical and subtropical part of the world. Chilli is known to be a host of about 23 begomoviruses resulting in serious economic losses to chilli growers. A majority of these viruses have emerged and cause serious problems in chilli cultivation and in last five years, the leaf curl disease of chilli has become the major constraint for chilli production in most of the chilli growing areas (Reddy *et al.*, 2005; Senanayake *et al.*, 2007). However, very little work has been done to diagnose the associated begomovirus (es) of leaf curl disease in India. There have been reports from Lucknow (Khan *et al.*, 2006a &b), Alipur (Reddy *et al.*, 2005) and Rajasthan (Senanayake *et al.*, 2007).

Therefore, keeping in view the economic importance of the leaf curl disease, an attempt was made to develop methods for the detection and identification of the causal agent on molecular level. We here present the data on symptomatology, disease incidence, biological studies, PCR detection and molecular characterization of begomovirus isolates causing leaf curl disease on chilli in northern India.

SYMPTOMATOLOGY AND DISEASE INCIDENCE

Symptoms of chilli leaf curl disease consisted abaxial and adaxial curling of the leaves accompanied with puckering and blistering of inter-venial areas, thickening and swelling of the veins, distortion of leaves, shortening of internodes, development of small branches and reduced growth of the plants with fewer deformed fruits (Fig 1). The incidence of severe leaf curl disease was observed high to moderate in various cultivars of chilli (*Capsicum annum*), during a survey conducted in various locations of Tarai region (Bahraivh) in Uttar Pradesh and in other parts of northern India (Lucknow and Kanpur). The infected plants of chilli were collected and maintained in glasshouse.

Fig 1. (a). A field view of chilli crop showing severity of chilli leaf curl disease and (b) a close view of a naturally infected chilli plant showing severe leaf curl and deformation of fruits.

BIOLOGICAL STUDIES OF BEGOMOVIRUS INFECTING CHILLI

The biological studies of chilli samples collected from Bahraich, Lucknow and Kanpur revealed that the virus of chilli leaf curl disease is efficiently transmissible through whiteflies in a persistent manner (each acquisition and inoculation period of 24 hrs) and induced symptoms in chilli similar to the naturally infected plants. The virus also produced characteristic symptoms mild to severe leaf curl in *Lycopersicon esculentum, Nicotiana tabacum* cv. White Burley and *N. glutinosa* but not on *Solanum melongena*. On the basis of positive whitefly transmission test in persistent manner, the virus (es) infecting chilli were considered to be members of whitefly-transmitted geminivirus (WTGs).

DETECTION BY POLYMERASE CHAIN REACTION (PCR)

Polymerase chain reaction (PCR) is an important tool for molecular diagnosis because of its high sensitivity and rapid detection. Initially two degenerate primers designed by Deng *et al*, (1994) were used during PCR for detection of begomovirus in total DNA isolated from chilli samples. Deng primers worked well and amplicon of expected size (~550 bp) was obtained which confirmed the association of begomovirus with chilli leaf curls disease.

Since the coat protein (CP) represents the most conserved gene in the family *Geminiviridae* which is informative as the entire sequence of the genome and is sufficient to classify a virus isolate. The CP nucleotide sequences/amino acid have been used to establish taxonomical relationship among WTGs by various workers. Therefore, the primers specific to CP (Srivastava *et al.*, 1995) and Replicase (Rojas *et al.*, 1993) genes of a well characterized begomovirus were employed. The PCR resulted in positive amplicons of ~800 bp, and ~1.1 kb respectively in all the chilli samples collected from Bahraich, Lucknow and Kanpur. Movement protein (Padidam *et al.*,1995) and DNA β specific (Briddon *et al.*, 2001) primers employed with chilli samples were also capable to amplify the appropriate size (~875 bp and 1.35 kbp) DNA fragments confirming the begomovirus contains the bipartite genome and a DNA beta molecule.

MOLECULAR CLONING AND SEQUENCE ANALYSIS OF BEGOMOVIRUS GENOME

Further, the causative pathogen was identified by cloning, sequencing and sequence analysis of the DNA-A component of begomovirus isolates. The complete DNA A of chilli begomovirus was amplified by PCR using a pair of specific primers of the DNA A of *Tomato leaf curl New Delhi virus* (ToLCNDV) (Fig 2a) and amplicon was cloned and sequenced. The analysis of sequence data of the virus isolate revealed the presence of seven ORFs: AC5, AV1 & AV2 in virion sence and AC1, AC2, AC3 & AC4 were in complementary sence of the genome (Fig 3a). The sequence analysis of the complete DNA A (Acc. EU309045) showed (92-97%) identities with several isolates of ToLCNDV reported from all over the world while it was found less than 90% with other tomato begomoviruses (Table 1). According to highest 97% identities and the criteria suggested by International Committee of Taxonomy of Viruses

(ICTV, Fauquet *et al.*, 2008) the virus from chilli has been identified as an isolate of ToLCNDV. The bi-partite nature of virus isolate was also confirmed by successful amplification and sequencing of the movement protein (MP) gene region of DNA-B component. On the basis of high identities and close phylogenetic relationships, the begomovirus associated with leaf curl disease of chilli was identified as a strain of ToLCNDV.

All the analysis based on nucleotide and amino acid sequence showed that there was a high degree of sequence identity at both the sequence levels among ToLCNDV isolates under study, and with various ToLCNDV strains reported from a variety of plant species. Following the criteria of ICTV report (Stanley *et al.*, 2005) and also the criteria proposed by Fauquet and Stanley (2005) and Fauquet *et al.*, 2008 based on maximum nucleotide sequence identity of CP chilli-Bahraich isolate was identified as the same virus (ToLCNDV-tomato), where as Lucknow and Kanpur isolates were identified as the variants of ToLCNDV-luffa and ToLCNDV-Mild, respectively of the ToLCNDV strains of *Begomovirus* genus.

Dendrogram alignment of ToLCNDV-chilli isolates (under study) at CP level clustered together with strains of ToLCNDV and fall in one group. Bahraich isolate showed close relationship with tomato-New Delhi (AY428769). Lucknow isolate clustered closely with tomato-Bangladesh (AJ875157) and chilli-Alipur (AJ810340), while Kanpur isolate showed most close relationship with tomato-mild India (U15016) (Fig 4). The phylogenetic tree based on Rep gene showed that all the ToLCNDV strains clustered together. A remarkable observation was that the Bahraich and Lucknow isolates were closest with each other than the Kanpur isolate. This feature was also noticed during Dendrogram analysis of coat protein gene. All the above observations clearly indicated that these ToLCNDV-chilli isolates under study are the strains of ToLCNDV.

The pair wise alignment of protein sequences (amino acid) of CP and Rep of ToLCNDV-chilli isolates (under study) with different ToLCNDV strains showed considerable substitution of amino acid throughout the sequence. The CP ORF of each ToLCNDV-chilli isolate was found most conserved than that of the Rep ORFs. The Rep ORFs showed a maximum substitution of amino acids. Out of these minor substitutions there were high degree of sequence identity among these chilli isolates with each other and with other ToLCNDV strains. The most considerable feature was noticed that Lucknow and Kanpur isolates are more closely related with each other at both CP and Rep level.

DNA Β MOLECULE ASSOCIATED WITH TOLCNDV-CHILLI ISOLATES

DNA β molecules are symptom modulating, single stranded circular DNA satellites associated with begomoviruses (family *Geminiviridae*). The molecules have been found associated with diverse plant species from India and neighboring countries. Thus, investigation of DNA β satellite molecule was also carried out to confirm the association of satellite molecule with ToLCNDV-chilli isolates under study. The total DNA isolated from the chilli samples was employed initially for dot blot hybridization with DNA β specific probe which showed strong signals of hybridization in all chilli samples (ToLCNDV). During PCR analysis, as expected to the primers, amplicons of approximately 1.35 kb were amplified in all three chilli samples (Fig 2b). These findings confirmed the association of the satellite molecule with the leaf curl disease.

Accession No.	Abbreviated virus name	Natural Host	Place/Country	Sequence identity (%)
AJ875157	To LCNDV	Tomato	Jessore, Bangladesh	97
AY428769	ToLCNDV	Tomato	New Delhi, India	97
EF035482	ToLCNDV	Okra	Rajsthan, India	94
AM286434	ToLCNDV	Pumpkin	New Delhi, India	94
AF448058	ToLCNDV	Tomato	Pakistan	94
Y16421	ToLCNDV	Tomato	Lucknow, India	96
DQ116880	ToLCNDV	Chilli	Khanewal, Pakistan	94
AM292302	ToLCNDV	Luffa	Multan, Pakistan	94
EF043230	ToLCNDV	Potato	Happur, India	93
EF063145	ToLCNDV	Cotton	Hisar, India	93
AB330079	ToLCNDV	Cucumber	Thailand	92
AF274349	ToLCSV	Tomato	Sri Lanka	63
NC_003897	ToLCKV	Tomato	Karnataka, India	66
DQ852623	ToLCBV	Tomato	Kerala, India	59
AF413671	ToLCGV	Tomato	Gujarat, India	65
DQ629103	ChLCV	Tomato	India	67

Table 1. Sequence identity of complete DNA A of *Tomato leaf curl New Delhi virus-chilli* (Bahraich EU309045) isolate based on Genomatix DiAlignment programme.

Abbreviations: ToLCNDV= *Tomato leaf curl New Delhi virus*, ToLCSV= *Tomato leaf curl Sri Lankan virus*, ToLCKV= *Tomato leaf curl Karanataka virus*, ToLCBV= *Tomato leaf curl Bangalore virus*, ToLCGV= *Tomato leaf curl Gujarat virus*, ChLCV=*Chilli leaf curl virus*.

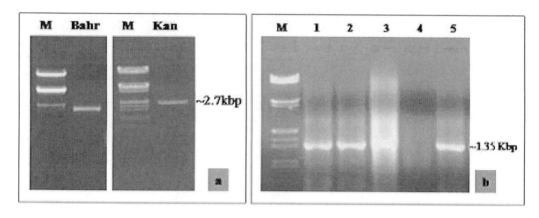

Fig. 2. (a). Positive amplicon of ~2.7 K bp from Bahraich isolate obtained was cloned, sequenced and deposited in GenBank: Acc.No. EU309045 as *Tomato leaf curl New Delhi virus*-Chilli pepper DNA A. (b). PCR amplification of ~1.35 kbp DNA fragments with DNA β specific primers. Lanes 1-3: Isolates of chilli from Bahraich, Lucknow and Kanpur respectively. Lanes 4-5: healthy and ToLCNDV infected tomato. M: *Eco* RI/*Hind* III digest lambda DNA marker.

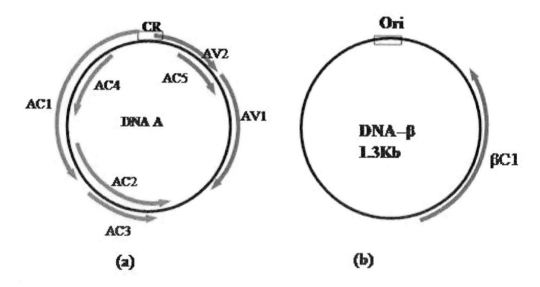

Fig. 3 (a) Genome organization of DNA A of Chilli begomovirus under study (Acc. EU309045) and (b) beta DNA associated with leaf curl disease of chilli (DQ343289).

Further, the amplicon obtained from ToLCNDV-chilli Bahraich isolate was cloned and sequenced which reveled the full length nucleotide sequence of 1371 bp. The data was submitted in GenBank (DQ343289). The Blast search analysis of nucleotide sequence revealed high level sequence identity with a number of DNA β components reported from a variety of plant species from different parts of the world The analysis of the nucleotide sequence of the DNA β associated with the ToLCNDV-chilli Bahraich isolate showed all the characteristics of DNA β viz. satellite conserved region (SCR), adenine rich region (A-rich region) and an open reading frame β CI (ORF β CI) (Fig 3b).

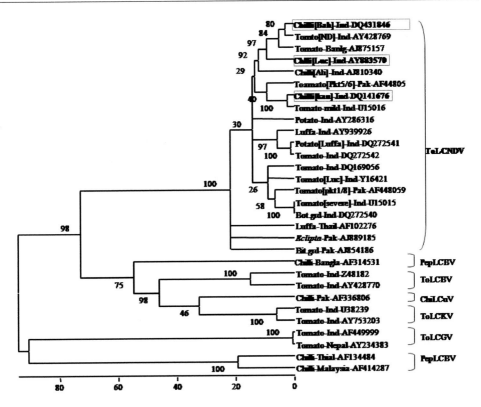

Fig 4. Phylogenetic relationship of ToLCNDV-Chilli isolates (under study) with other selected begomoviruses based on the alignment of nucleotide sequences of coat protein (AV1) gene.

MANAGEMENT STRATEGIES OF THE CHILLI LEAF CURL DISEASE/VIRUS

As whitefly (*Bemisia tabaci*) is the efficient vector of the begomoviruses, a common insecticide 'Malathion' was used to minimize the populations of insect-vector (*B. tabaci*). The three round foliar spraying of 0.2% insecticide after 21 days interval was found promising for the control of whitefly in chilli crop. Improvement in plant growth and appreciable enhancement in fruit yields was also recorded in sprayed plants of three chilli cultivars as compared to the unsprayed plants. Therefore, three sprayings of Malathion insecticide (0.2%) at 21 days intervals is recommended for possible management of leaf curl disease of chilli (Khan *et al.*, 2006b). Since chilli fruits are used as raw and in food preparations and use of insecticides is known to cause adverse effect on human health therefore, spraying of insecticide on chilli should be avoided. Therefore an alternate method of disease management may be the best choice.

Neither an effective control measure of begomoviruses nor resistant varieties of chilli have been developed in our country. Keeping this in mind the development of regeneration procedure of chilli was attempted which is a pre-requisite for the rasing of transgenic plant. *Capsicum annuum* cv. Pusa Jwala cotyledons (leaf disk) and hypocotyls (internodal segments) explants were cultured on MS basal medium supplemented with different combination and concentration of auxin and cytokinin. The medium supplemented with 5mg/l

BAP (cytokinin) and 0.1mg/l NAA (auxin) was found optimum for the development of shoots from cotyledon explants while the supplementation of 10mg/l BAP and 0.1mg/l NAA was appropriate for hypocotyl explants (Fig 5).

Fig. 5. Various stages of regeneration of chilli cv. Pusa Jwala: *in vitro* grown seedlings (a), leaf discs on MS medium (b), internodal segments on MS medium and initiation of shoot buds (c), shoots transferred to jars for elongation (d), shoots in rooting medium showing roots (e), acclimatization of regenerated shoots (f), fully matured and well established plants with flowers and fruits (g-h).

The regeneration protocol of cv. Pusa Jwala developed may be utilized for the transformation purpose for virus-derived resistance (VDR) against ToLCNDV. The coat protein, rep protein and movement protein genes of ToLCNDV isolate characterized during this study may be utilized for their mobilization into the genome of chilli for the development of inbuilt resistance against the leaf curl disease caused by ToLCNDV strains.

REFERENCES

[1]   Anonymous. (2005). Agricultural News. *National Academy of Agricultural Sciences,* Pusa New Delhi. Volume XI Issue No. 2

[2]   Briddon R.W., Mansoor S., Bedford I.D., Pinner M.S., Saunders K., Stanley J., Zafar Y., Malik K.A. & Markahn P.G. (2001). Identification of DNA components required for induction of cotton leaf curl disease. *Virology* 285: 234-243.

[3]   Deng, D., Mc Grath, P.F., Robinson, D.J. & Harrison, B.D. (1994). Detection and differentiation of whitefly transmitted geminivirus in plants and vector insect by the polymerase chain reaction with degenerate primers. *Ann. Appl. Biol.* 125: 327-336.

[4]   Fauquet, C.M. & Stanley, J. (2005). Revising the way we conceive and name viruses below the species level: A review of geminivirus taxonomy calls for new standardized isolates descriptors. *Archives of Virology* 150: 2151-2179.

[5]   Fauquet, C.M., Briddon, R.W., Brown, J.K., Moriones, E., Stanley, J., Zerbini, M. & Zhou, X. (2008). Geminivirus strain demarcation and nomenclature. *Archives of Virology,* 153: 783-821.

[6]   Khan, M.S., Raj, S.K. & Singh, R. (2006 a). First report of Tomato leaf curl New Delhi virus infecting chilli (*Capsicum annuum*) in India. *Plant Pathology* 55: 289.

[7]   Khan, M.S., Raj, S.K., Bano, T. & Garg, V.K. (2006 b). Incidence and management of mosaic and leaf curl diseases in cultivars of chilli (Capsicum annuum). J. *Food Agriculture and Environment* 4: 171-174.

[8]   Padidam, M., Fauquet, C.M. & Beachy, R.N. (1995). Tomato leaf curl geminivirus from India has a bipartite genome and coat protein is not essential for infectivity. *Journal of General Virology* 76: 25-35.

[9]   Reddy, C.R.V., Colvin, J., Muniyappa, V. & Seal, S. (2005). Diversity and distribution of begomoviruses infecting tomato in India. *Archives of Virology* 150: 845-867.

[10]   Rojas, M.R., Gilbertson, R.L., Russel, D.R. & Maxwell, D.P. (1993). Use of degenerate primers in the polymerase chain reaction to detect whitefly transmitted geminiviruses. *Plant Disease* 77: 340-347.

[11]   Senanayake, D.M.J.B., Mandal, B., Lodha, S. & Varma, A. (2007). First report of Chilli leaf curl virus affecting chilli in India. *Plant Pathology* 56:343.

[12]   Srivastava, K.M., Hallan, V., Raizada, R.K., Chandra, G., Singh, B.P. & Sane P.V. (1995). Molecular cloning of Indian tomato leaf curl virus genome following a simple method of concentrating the supercoiled replicative form of viral DNA. *Journal of Virological Methods* 51: 297-304.

[13]   Stanley, J., Bisaro, D.M., Briddon, R.W., Brown, J.K., Fauquet, C.M., Harrison, B.D., Rybicki, E.P. & Stenger, D.C. (2005). *Geminiviridae.* In: Fauquet, C.M., Mayo, M.A. Maniloff, J. Desselberger, U. and Ball, L.A. (eds), Virus Taxonomy, VIIIth Report of the ICTV. Elsevier/Academic Press, London, pp 301-326.

In: Recent Trends in Biotechnology and Microbiology
Editors: Rajarshi Kumar Gaur et al., pp. 65-79

ISBN: 978-1-60876-666-6
2010 Nova Science Publishers, Inc.

Chapter 6

VIRUS (ES) INFECTING STONE FRUITS: DETECTION METHODS AND BIOTECHNOLOGICAL APPROACHES TOWARDS MANAGEMENT STRATEGIES

Vanita Chandel, Tanuja Rana, Vipin Hallan and Aijaz A. Zaidi*

Plant Virology Lab, Institute of Himalayan Bioresource Technology (CSIR), Palampur-176061, HP. India

ABSTRACT

Stone fruits includes almond, apricot, cherry, nectarine, peach and plum and these are grown on commercial scale in the North and North-Eastern States of India. Losses in terms of quality and quantity are reported from all over the world due the viral infections in these fruits. Important viruses infecting these fruits are: *Apple mosaic virus* (ApMV), *Prunus necrotic ringspot virus* (PNRSV), *Prune dwarf virus* (PDV), *American plum line pattern virus* (APLPV), *Plum pox virus* (PPV), *Apple chlorotic leaf spot virus* (ACLSV), *Apricot latent virus* (ApLV) and Plum bark necrosis stem pitting associated virus (PBNSPaV). In this chapter major virus infecting stone fruits has been described i.e. *Prunus necrotic ring spot virus*. Latest biotechnological approaches used in the management practices like virus free production of plants, CP mediated resistance and RNAi used to manage viral diseases are also described.

INTRODUCTION

Fruits are the means by which flowering plants disseminate seeds. In cuisine, when discussing fruit as food, the term usually refers to those plant fruits that are sweet and fleshy, examples of which include plums, apples and oranges. Many hundreds of fruits, including fleshy fruits like apple, peach, pear, kiwifruit, watermelon and mango are commercially valuable as human food, eaten both fresh and as processed food like jams, marmalade and other preserves. Fruits are also in manufactured foods like cookies, muffins, yoghurt, ice

---
* corresponding author e-mail- vanita_chandel@yahoo.co.in

cream, cakes, and many more. Many fruits are used to make beverages, such as fruit juices (orange juice, apple juice, grape juice, etc) or alcoholic beverages, such as wine or brandy.

Top ten countries in fruit production include China, India, Brazil, USA, Italy, Spain Mexico, Iran, Philippines and France. India contributes about 10 per cent of the world production of fruits thus surfacing as the second largest producer of the same. The annual production of the fruits in India is 41 million tones (http://www.fao.org/). Fruits as a group occupy an area of 0.11 million ha, with 0.14 million tons production. Apples and pears are the most important temperate pome fruit trees cultivated in India, with 227679 ha and 38539 ha, respectively, followed by stone fruits with 110000 ha (Ghosh, 2001). Introduced cultivars are widely cultivated but national breeding programs are now generating improved varieties with good adaptation to local condition and quality. Even though there are large differences among growers, most of them use old traditional practices and yields are relatively low. Research and extension are providing support to increase production efficiency.

In Himachal Pradesh about 28 thousand ha covered under stone fruits viz. peach, plum, apricot, almond and cherry. In North-western Himalayan region, peach holds greater promise because of its utilization for canning purposes. Peach is grown mainly in low and mid hilly areas (1000-2000 m a msl), except the low chilling cultivars belonging to the Florida group, which can be grown very well under sub-tropical conditions.

In the North-Eastern Hills (NEH) region, these crops can be successfully grown in Arunachal Pradesh (Kameng, Siang, Tirap and Lohit districts), Meghalaya (Central plateau of Khasi & Jaintia hills), Manipur (Maram, Tadubi, Mou, Ukhrul) and Nagaland (Mokokchung, Wokhla, Thensang, Kohima, Phek districts). The productivity of all the stone fruits is low and estimated yield of peach is 0.50, 0.73, 2.11 and 2.23 t/ha in HP, J&K, UP, and NEH regions respectively. The productivity of cherry in J&K is approximately 1.73 t/ha, while average yield of apricot is 0.42, 0.20 and 0.28 t/ha in H.P, J&K and U.P. hills respectively (http://www.fao.org/).

In India Horticultural exports contribute about 8 per cent in total agricultural exports this calls for efficient production and marketing efforts. Significant development has taken place in international trade regime for fair play and non-discriminatory trade. This has bearing for Indian economy. Many countries are interested in Indian market, because of market access provision in the WTO regime. The increasing consumption of fruits coupled with increase in income will certainly help such nations to export until India takes remedial measures to increase productivity and improve quality of the products. Therefore, many efforts are needed to exploit the world market potential especially for temperate fresh fruits when India have tough competition in fruits from USA, France, Belgium, Australia, Italy, Spain etc (Atteri, 2004).

Stone and pome fruits are infected by wide range of pathogens such as fungi, bacteria, virus, viroid and phytoplasma. Some of the important pathogens of these fruits are: peach leaf curl, brown rot, scab, mushroom root rots, canker, bacterial spots of leaves and fruits and crown gall are some of the fungal and bacterial diseases. Viral pathogens includes Prunus necrotic ring spot virus (PNRSV), Apple mosaic virus (ApMV), Apple chlorotic leaf spot virus (ACLSV), Prune dwarf virus (PDV), Plum pox virus (PPV) and viroids such as Peach latent mosaic viroid, Hop stunt viroid , Apple scar skin viroid and Pear blister canker viroid.

Prunus necrotic ring spot virus is one of the important viruses infecting almost all stone fruits and pome fruits. Different reports from all over the world shows the losses caused by this virus on stone fruits (Wood and Fry, 1984; Bech (1991); Ramsdell *et al.* (1992); Kajati and Szonyegi (1996); Hilsendegen (1999).

Table 1. Stone fruits area under cultivation and production in the world and India.

Fruit	Area under cultivation ha (World)	Total Production (Mt.)	Area under cultivation (ha)(India)	Total Production (Mt)
Peach	1,430,612	15,673,847	18,500	150,000
Plum	2,455,414	9,843,398	14,000	80,000
Apricot	434,581	2,821,223	2,400	10,000
cherry	406,594	1,858,673	1,700	8,000

Economic importance of the stone fruits for the hilly states of India is due to suitable climate conditions. However it is overcome by the fact that plant viruses infecting these trees sharply decrease their production. The major problem is at the level of selecting disease free planting material for raising healthy orchards. The problem is compounded by the fact that a thorough screening of these crops has not been done from the aspect of virus infections in Indian scenario. Therefore not much information is available and only limited reports are available where viruses infecting these crops have been identified mainly on the basis of symptomatology, transmission and serological detection (Bhardwaj and Thakur, 1993).

LOSSES CAUSED BY VIRUSES IN FRUIT CROPS

Viral diseases cause economic losses through lower yields and reduced quality of plant products. Viral diseases in perennial crop plants are more dangerous than in annual crops. Viruses can remain latent, spreading through an orchard and inflicting damage. Latent infestations can produce small to moderate losses in fruit production (Agrios, 1997). The literature reports different types of damages in fruit products, which includes unmarketable fruits (Reeves and Cheney, 1962), substantial reduction in yields and extensive tree death (Stouffer and Smith, 1971).Overall viral infections have greater effect on crop yields and fruit quality (deformation and loss of flavors) than on vegetative growth. With the most virulent strains the yield losses can reach 98% (Nemeth, 1986). *Prunus necrotic ringspot virus* is responsible for growth and reductions that vary from 12% to 33% (Pine, 1964). Some viral infections cause incompatibility between rootstock and cultivar affecting yields of nurseries. A low percentage of successfully grafted trees in nurseries may be attributed to viral infection (Nemeth, 1986). Mixed infections of PNRSV, PDV and ACLSV caused decline, low yields and unusual fruit disorders in apricot (Sutic *et al.*, 1999). Cembali and group (2003) gave the economic implications of a virus prevention program in deciduous tree fruits. They gave the economic impact of viral diseases on fruit trees and losses caused by them on the fruit production and to the fruit industry.

Symptoms Caused by PNRSV

Symptoms of PNRSV are characterized by discoloration or localized necrosis of foliar tissues, flowers (color breaking of peach flowers), buds (bud failure of almond, peach mule's ear), and shoots (terminal die-back). Discoloration in the form of chlorotic or bright yellow blotching (almond calico), chlorotic to yellow mottling, ringspots, line and oak leaf patterns particularly in rose, diffuse yellowing (cherry dusty yellows), are especially conspicuous in spring on newly developed vegetation, but tend to fade as the season advances (Brunt et al., 1990).

Different viral symptoms were observed during the surveys in the present study. Typical viral symptoms such as mosaic, cholorsis, shot holes, necrotic spots, pox were observed on various stone and pome fruits (Fig.1)

Detection of PNRSV

ELISA: After the discovery of enzyme linked immunosorbent assay (ELISA) by Clark and Adams (1977) as the tool for the detection of virus, other techniques of plant virus detection like biological indexing, agar double diffusion tests, liquid precipitin tests, latex flocculation tests have become outdated because of the high sensitivity and reproducibility of ELISA over other detection methods. Thereafter ELISA has been used extensively and is now the most preferred method for the routine diagnostic and disease incidence purposes for plant viruses (Salem et al., 2003b). The minimum concentration of PNRSV detectable by the latex test was 800 ng/ml, by ELISA 4 ng/ml and by SSEM 4 ng/ml. In 1980, Thomas used purified virus preparation of ArMV, PNRSV and SLRSV to assess the sensitivities of 4 serological methods viz. ELISA, immunodiffusion in gels, latex flocculation assay and SSEM, for the detection of these viruses. The latex test was found to 250 times more sensitive than gel immunodiffusion, but SSEM and ELISA were, respectively, up to 1000 and 200 times more sensitive than the latex test. Later on, Kelley and Cameron (1986) used ELISA for the detection of PNRSV in sweet cherry and almond seeds, respectively.

ELISA has been used to detect the virus in suspected samples. Commercial kits available in the market for the detection of PNRSV were used (Agdia). Positive results were obtained in almond, cherry, plum, peach and apple (Chandel et al., 2007a, 2007b, 2008a and 2008b). Similarly another virus i.e. Apple cholorotic leaf spot virus was also detected in most of the stone fruits (2007b, 2008a, 2008b, 2008c, 2009) and in apple as a major virus (Rana et al., 2007a).

Hybridization of the total RNA/viral genomic RNA from the infected plant species with the radiolabeled or non-radiolabeled virus specific DNA/RNA probe is a very powerful tool for the detection and identification of virus from the infected plants, because of its very high efficiency and sensitivity. It can also be used for serotype differentiation. Herranz and co-workers in 2005 used molecular hybridization technique for the detection of stone fruit viruses. The 'polyprobe four' (poly4) covered the four ilarviruses affecting stone fruit trees including Apple mosaic virus (ApMV), Prunus necrotic ringspot virus (PNRSV), Prune dwarf virus (PDV), and American plum line pattern virus (APLPV).

Figure 1. Symptoms observed on plum and nectarine in some of the orchards which were surveyed. A) & B) Shows mosaic and shot holes symptoms on plum. C) Shows mosaic and leaf deformation on plum. D) Shows chlorosis in plum. E) Necrotic spots in cherry leaves. F) Pox like symptoms on nectarine. G). Necrotic spots on nectarine.

RT-PCR: RT-PCR has been used extensively for the detection of plant viruses. The technique is very sensitive and is able to detect even picograms of virus. As the technology progress RT-PCR also getting advanced day by day so that it can detect even smaller amount of viruses in the sample more reliably and efficiently

Spiegel and co-workers (1996) observed that RT-PCR technique could be used for detection of PNRSV in dormant peach trees which were earlier tested negative by ELISA. They further observed that, in some cases, PCR products were not clearly visible in the stained gel and became distinct only after hybridization with a P ^{32}-labeled virus specific probe. Rosner and group (1997) improved the detection of PNRSV in dormant peach and almond trees by the application of 2 different pairs of primers yielding a short and a long product, respectively. A new and sensitive colourimetric-PCR assay for detection of viruses in woody plants has been developed by Rowhani and co-workers in (1998) which did not rely on gel electrophoresis or molecular hybridization to detect virus-specific PCR products. This colourimetric method for detection of PNRSV from woody plants was demonstrated to be 3 orders of magnitude more sensitive than gel analysis. To further increase the sensitivity, this technique could be combined with immunocapture of virions from plant sap. Rosner and his group in (1998) investigated the reaction conditions for the detection of PNRSV in peach trees using IC-RT-PCR. They found that incubation of the reverse transcriptase reaction at 46°C resulted in higher levels of amplified products as compared with the recommended 37°C. Preheating the reaction mixture at 55°C for 5 min further improved PCR yields.

RT-PCR was used to amplify CP gene of PNRSV in samples which were found positive by ELISA and desired amplification of 700 bp was obtained in gel electrophoresis. The gene was further cloned and sequenced using standard protocols (Chandel *et al.*, 2007a, 2007, 2008a, 2008b).

MANAGEMENT STRATEGIES PRODUCTION OF VIRUS FREE PLANTING MATERIAL BY TISSUE CULTURE TECHNIQUES

Losses caused by viruses are both qualitative and quantitative. Unlike fungus and bacteria it is difficult to manage viruses by chemical control and conventional methods used for other pathogens and Nemeth in 1986 demonstrated the detrimental effect of the PNRSV and ChRSV on the development of 2-3 year old sour cherry trees. PNRSV reduced tree height by 2-8%, the length of the shoots by 2-22% and trunk diameter by 5-27%. Detrimental effects of virus infections on growth and yield is also reported on apple trees. ApMV can reduce plant growth by 50%, trunk diameter by 20% and fruit yield by 30% in susceptible cultivars (Posnette and Cropley, 1959). Ringspot viruses are particularly damaging in stone fruit production. PNRSV may reduce the sour cherry yields by 91-98% (Sutic *et al.*, 1999). Such high losses characterize the acute stage of the disease, while losses of 36-43% occur in chronic stage (Kunze, 1969). PNRSV and PDV are shown to reduce fruit set by 37% in four peach cultivars (Topchiiska, 1983). So after having a look on the losses caused by viruses it is important to make them virus free to prevent losses.

Cultivation of virus resistant genotypes are often restricted by the availability of a limited germplasm pool for a source of virus resistant as opposed to the wide host range of viruses in addition breeding programme is laborious and time consuming, especially for the crops with long generation cycles. Consequently the most effective measure available so far for the virus eradication is use of virus free propagation material obtained by employing the meristem tip culture alone or in combination with chemo and thermotherapy. Engel (1990) studied the importance of virus free planting material in integrated fruit production by observing the behavior of virus free and virus infected apple and pear trees and found that the virus free trees were more vigorous, had higher fruit yields and better quality fruits than virus infected trees and they did not required additional treatments with fertilizers and growth regulators. They also found that virus free trees were also better for successive replanting.

The technique virus elimination pioneered by Morel and Martin (1952) was based on the assumption that virus particles are unevenly distributed in the plants and their titer decreases as the meristmatic dome of vegetative shoot apex is approached. The apical meristem domes were thus considered free from virus particles. Virus free production of plants is necessary from the point of view to prevent further spreading of virus. Use of virus free propagative material will prevent further dissemination of viruses in the field and thus can prevent long term losses.

Virus elimination by meristem tip culture and tip micrografting combined with the use of chemotherapy and thermotherapy is mentioned by Faccioli and Marani (1988). Mainardi and group in 1992 used DAS-ELISA to test shoot tip cultures of Mr.S. 2/5 plum rootstock for *Prunus necrotic ringspot ilarvirus* (PNRSV) and *Plum pox potyvirus* (PPV). Micropropagated shoot tips subjected to intensive sub culturing showed a higher rate of infection and were more reliable than screen house grown plants as source material in testing for the presence of both viruses. All the plantlets derived from *in vitro* propagation retained the virus infection, suggesting that PNRSV and PPV in Mr.S. 2/5 plum rootstock cannot be eradicated simply by intensive *in vitro* subcultures of shoot tips. It is concluded that the use of *in vitro* propagated material for indexing can bring some advantages to programmes for the certification of the fruit trees. Eremin and Podorozhnyi in 1992 produced healthy planting material of cherry, plum and stone fruit rootstocks by tissue culture.

Studies by Gabova, 1995, *in vitro* propagated plums (*P. domestica*) systemically infected with *Plum pox potyvirus* were used as genetic material. The viricide ribavirin (virazol) was incorporated in nutrient media in which the 4 varieties were cultured. After 30 days' treatment, shoots 2 cm long were placed in ribavirin-free medium. Almost all virus/genotype combinations responded with virus titre inhibition, in the susceptible Pozegaca and Tetevenska this inhibition was 14-33% while in the tolerant Stanley and Opal it was 32-61%. Treatment with 10 mg/litre inhibited virus titre by 75-91% in Stanley and Opal and by 30-46% in the susceptible varieties. The ability of Opal to localize the infection (hypersensitive response) under field conditions seemed to occur in culture also, especially at 10 mg/litre (of 3 doses tested). Pozegaca maintained a fairly high rate of proliferation when treated with ribavirin.

Knapp and co-workers in 1995 carried out *in vitro* virus elimination programme for *Malus* and *Prunus* spp. using *in vitro* thermotherapy and meristem tip culture. Zotto and Docampo in 1997 standardized the conditions for the establishment of *in vitro* cultures of Marianna 2624 and Pixy plum rootstocks free from *Prunus necrotic ringspot ilarvirus*, *Prune dwarf ilarvirus, Plum line pattern ilarvirus*, phytoplasmas and *Xylella fastidiosa*.

Elimination of *Sugarcane yellow leaf mosaic virus* by meristem tip culture is reported from Germany (Fitch *et al.*, 2001). Meristem tip culture is reported for the production of virus free *Prunus* sp. by Boxus in 1975. PPV and PNRSV free nectarine plants have been produced by thermotherapy and meristem tip culture (Manganaris *et al.*, 2003). Similarly banana viruses were eradicated by *in vitro* tissue culture methods (Hazaa *et al.*, 2006). From ornamentals also virus has been successfully eradicated using tissue culture techniques. In begonias PNRSV was eliminated by using meristem tip culture and chemotherapy (Verma *et al.*, 2005). Most recently Golino and co-workers (2007) eliminated the rose mosaic virus using micro shoot tip tissue culture. Virus free chrysanthemums were produced by meristem tip culture, chemotherapy and thermotherapy techniques (Verma *et al.*, 2007). Production of Indian *Citrus ring spot virus* free plants of Kinnow employing chemotherapy coupled with shoot tip grafting is reported from India (Sharma *et al.*, 2008). Thermotherapy and chemotherapy *in vitro* were applied to eliminate *Apple chlorotic leaf spot virus* (ACLSV) and *Prunus necrotic ring spot virus* (PNRSV) from myrobalan (*Prunus cerasifera* var. *divaricata* Borgh), PNRSV from 'Empress' plum (*P. domestica* (L.) and *Prune dwarf virus* (PDV) from 'Early Rivers' sweet cherry (*Cerasus avium* (L.) plants. ELISA assays conducted one year after therapy indicated that thermotherapy *in vitro* was highly effective for PNRSV and ACLSV elimination (Cieślińska, 2007). Engel in 1990 gave the importance of virus-free plant material in integrated fruit production trees and found that virus free trees produced by tissue culture were more vigorous, had higher fruit yields and better quality fruit than virus-infected trees, and did not require additional treatments with N-fertilizers and growth regulators. Virus-free trees were also better for successive replanting, and could be planted at lower densities than virus-infected trees. Wang and group (2000) studied the reaction of virus-free apple tree growth and fruit production and studied the growth and fruit production of virus-free apple cv. Fuji, Gold Spur, Tian Huangku and Ralls for 8 years. Compared with control trees, the height, crown and trunk diameter of virus-free trees increased by 6.4 to 16, 9.8 to 15.9 and 8.9 to 17.4%, respectively, while the number of new branches and total new branch length per tree increased by 12.7 to 24.4% and 23.1 to 40.6%, respectively. The fruit bearing of biennial virus-free Gold Spur, triennial virus-free Fuji, and triennial and 4-year-old virus-free Ralls reached 60, 48, 62.5 and 97.5% respectively, while the corresponding virus-infected trees reached 16.7, 16, 0 and 12.5%, respectively. The number of fruits in the virus-free Tian Huangkui, Gold Spur and Fuji was higher by 41.6, 54.7 and 45%, respectively, than those of the virus-infected trees. The fruits of virus-free Fuji, Gold Spur and Ralls were 8.8-22.3% harder than those of the virus-infected fruits. The ascorbic acid, total sugar and acid, and anthocyanin in fruits of virus-free trees were higher than those of the virus-infected trees. Wei and Timon (1994) studied the improvement of efficiency in producing virus-free peach (*Prunus persica*) plants *in vitro*. Vertesy (1980) studied *in vitro* propagation of *Prunus persica* and *P. persico-davidiana* shoot tips in order to get virus-free plants.

Yamaga and Munakata (1991) produced virus-free apple planting stock by meristem culture. Bhardwaj and group (1998) used meristem tip culture and heat therapy for production of *Apple mosaic virus* free plants in India.

Laimer and co-workers in 2006 eliminated the viruses of *Prunus* spp. by tissue culture techniques. Rizqi and group in 2001 produced virus (PNRSV) free almond plants by shoot tip grafting *in vitro*. Paunovic and co-workers in 2007 obtained PPV free plums by chemotherapy with ribavirin *in vitro*. An *in vitro* virus elimination programme for *Malus* and *Prunus* spp. was carried out using *in vitro* thermotherapy and meristem tip culture. After the treatments, *Malus* sp. was tested for *Apple mosaic ilarvirus* (ApMV), *Apple stem grooving capillovirus* (ASGV) and *Apple chlorotic leaf spot trichovirus* (ACLSV), and Prunus for *Plum pox potyvirus* (PPV), *Prunus necrotic ringspot ilarvirus* (PNRV), *Prune dwarf ilarvirus* (PDV), ApMV and ACLSV by ELISA and immuno tissue-printing. PPV, PNRSV, ASGV and ACLSV were detectable after 4 years of *in vitro* culture. PDV was found in an 8-month old Prunus culture. ApMV was lost after 2.5 years of *in vitro* culture. PPV could be detected in leaves after 20 d of thermotherapy, but not after meristem culture. ASGV was either positive after 33 d of thermotherapy or reduced below the threshold level for reliable serological detection. ASGV titre significantly increased 6 months after thermotherapy and meristem dissection. No ACLSV positives were found after 33 d of thermotherapy, and shoots grown from meristems of main shoots were free of ACLSV 6 months after, whereas shoots from axillary meristems were infected (Knapp *et al.,* 1995).

One of the major problems of plum growing is related to the viral disease *Plum Pox virus* (PPV). It is a limiting factor for spreading a number of plum cultivars with valuable economic and biological characteristics. On the other hand, the *in vitro* method enabled in obtaining of great quantities of virus-free planting material for a shorter time, which was difficult to be achieved by the conventional methods. Kornova and his group in 2002 work out the technological parameters for the production of PPV-free planting material of 4 plum cultivars on the basis of *in vitro* propagation and virology control. The plum cultivars were selected according to their economic significance and the relative resistance to PPV, cvs. 'Cacanska lepotica' and 'Althan's gage', belonging to *Prunus domestica* L. species and cvs. 'Santa Rosa' and 'Black Star' of *Prunus salicina* L. species. The parameters of introducing the explants into sterile culture, the nutrient media and the conditions of multiplication and rooting and the adapting under *in vivo* conditions have been detected. In the process of studies a virology control of the source mother plants marked for explants, as well as of the obtained micropropagated and adapted plum plants has been conducted (Kornova *et al.*, 2002).

PPV-free plants of *Prunus domestica* cv. 'Kyustendilska Sinya' (KS) and cv. 'Valjevka' (Val) via *in vitro* techniques (Nacheva *et al.*, 2002).

In the present study attempt have been made to make Prunus plants virus free using meristem tip (0.2-1.0mm) culture and chemotherapy. For chemotherapy two chemicals viz. 2-thiouracil and virazole were standardized by using at different concentrations in the growth medium. In Mayfair variety 2-thiouracil at conc. 50mg/l gave maximum virus free plants i.e. 76.83% from ELISA and 66.66% from RT-PCR. Similar results were obtained in Snow queen variety; maximum virus free plants were obtained at same concentration i.e. 79.66% by ELISA and 66.66% by RT-PCR. Ribavirin gave best results at conc. of 30µg/lt, giving 83.33 % virus free plants when tested by ELISA and similar percentage when tested by RT-PCR in Mayfair.

OTHER BIOTECHNOLOGICAL APPROACHES

CP Mediated Resistance

CP mediated resistance was obtained against Plum pox virus in apricot. Transgenic plants were obtained by co cultivation with *Agrobacterium tumefaciens* strain containing various binary plasmids and coat protein gene of *Plum pox virus* (Machado *et al.*, 1993). Similarly Scorza *et al.*, 1994 produced transgenic plums by expressing the PPV CP gene. Similar studies were done by Lopez-Moya *et al.*, 2000, to make plants PPV resistance by expression of PPV CP gene in transgenic plants. Revalonandro and co-workers in 2000 also used CP mediated resistance strategy against PPV infecting plum.

Rnai: Approach to Achieve Virus Resistant Plants

As a spin-off of initially unexplained effects observed in protein mediated resistance approaches, the role of RNA transcripts of viral transgenic made a major contribution to the discovery of an entirely new field in biology involving sequence-specific RNA breakdown. This post-transcriptional gene silencing (PTGS) process, occurring in plants but also in other eukaryotes, is also known as RNA interference (RNAi) or RNA silencing.

If we think about the management strategies the most effective used these days is the RNAi mechanism to control plant viruses. RNA interference is a post-transcriptional gene regulatory mechanism. RNAi is induced by double stranded RNA and is mainly involved in defense against viruses. To counteract RNAi viruses have RNAi suppressors. RNAi uses small double-stranded RNAs to silence genes bearing a complementary sequence. Although endogenous gene silencing operates through multiple mechanisms, including mRNA cleavage, inhibition of translation, and epigenetic modifications of chromatin, mRNA cleavage is the most efficient mechanism and is the mechanism being harnessed for antiviral therapies. Small RNAs, either exogenous small interfering RNAs (siRNAs) or endogenous microRNAs, are taken up by a cytoplasmic RNA-induced silencing complex (RISC), which cleaves one strand, leaving the remaining unpaired guide strand to search for mRNAs bearing complementary sequences. Once recognized, if the target site on the mRNA has nearly perfect complementarity to the guide siRNA, the mRNA is cut by an Argonaute endonuclease in the RISC and then degraded, silencing the expression of the protein it encodes.

Damiano and his co-workers in 2007 used same strategy against PPV they transformed *Nicotiana benthamiana* plants with four PPV sequences, covering the P1 and HC-Pro genes, arranged to express self-complementary "hairpin" RNA. Two lines for each construct were challenged with PPV. All the transgenic plants were resistant to PPV infection.

Experiments have proved RNAi technology to be suitable to control infection by a variety of plant viruses. PPV is the causal agent of sharka or plum pox, the most serious disease on woody trees of the genus *Prunus* (Tenllado *et al.*, 2004).

CONCLUSIONS

The climatic conditions of India are well suited for the production of stone fruits predominantly in Himachal Pradesh, Jammu & Kashmir, Uttrakhand and some parts of Eastern States. But their production is largely affected due to the viral infection. A number of viruses are known to infect various stone fruits and the data on the losses caused by these viruses are available. During surveys in several stone fruit growing places in India many viruses were found but the incidence of *Prunus necrotic ring spot virus* (PNRSV) was more than other viruses. In the management practices various techniques such as virus free plant production by tissue culture, CP-mediated resistance and RNAi as a major emerging tool can be used effectively to manage the viruses

ACKNOWLEDGMENTS

The authors are thankful to the Director IHBT CSIR, Palampur for providing the facilities to carry out the research work and to the Department of Science and Technology to provide funding under WOS-A Scheme.

REFERENCES

[1]   Agrios, G.H. (1997). Plant Pathology. 4th Edition, San Diego, Academic Press, 635pp.
[2]   Atteri, B.R. (2004). Temperate fruit production domestic marketing and international trade issues.*Acta Horticulturae,* 662: 459-463.

[3]   Bech, K. (1991). *Prunus necrotic ringspot virus* (PNRSV) in sour cherry. Symptoms, incidence in orchards and influence on fruit yield. *Tidsskrift for Planteavl,* 95: 223-232.

[4]   Bhardwaj, S.V. Rai, S.J. Thakur, P.D. & Handa, A. (1998). Meristem tip culture and heat therapy for production of *Apple mosaic virus* free plants in India. *Acta-Horticulturae.,* 472: 135-140.

[5]   Bhardwaj, S.V. & Thakur, P.D. (1993). Virus diseases of temperate fruits. In: Chadha, KL and Pareek, OP. Advances in Horticulture. Vol. 3 - Fruit Crops: Part 3. New Delhi, Malhotra Publishing House, 1993, pp. 1447-1452.

[6]   Boxus, P. (1975). Meristem tip culture for the production of virus frees Prunus. *Acta Horticulturae.,* 44:43-46.

[7]   Brunt, A., Crabtree, K. & Gibbs, A. (1996). Viruses of tropical plants. Wallingford, UK, CAB International.

[8]   Cembali, T., Folwell, R.J., Wandschneider, P., Eastwell, K.C. & Howell, W.E. (2003). Economic implications of a virus prevention program in deciduous tree fruits in the US. *Crop Protection.,* 22 : 1149-1156.

[9]   Chandel, V., Rana, T. Hallan, V. & Zaidi. A. A (2008b). Occurrence of *Prunus necrotic ring spot virus* on nectarine (*Prunus persica*) in India. *Bulletin OEPP/EPPO Bulletin.,* 38: 223–225.

[10]   Chandel, V., Rana, T., Hallan, V. & Zaidi, A.A. (2007a). Evidence for the occurrence of *Prunus necrotic ring spot virus* on Peach in India by Serological and Molecular Methods. *Canadian Journal of Plant Pathology,* 29:311-316.

[11]   Chandel, V., Rana, T., Hallan, V. & Zaidi, A.A. (2007b). Wild Himalayan Cherry (*Prunus cerasoides*) as natural host of *Prunus necrotic ring spot virus* in India. *Plant Disease,* 91: 1686-1686.

[12]   Chandel, V., Rana, T., Handa, A., Thakur, P.D., .Hallan, V. & Zaidi. A.A. (2008a). Incidence of *Prunus necrotic ring spot virus* on *Malus domestica* in India.*Journal of Phyotopathology,* 156:382-384.

[13]   Cieślińska, M. (2007). Application of thermo and chemo therapy *in vitro* for eliminating some viruses infecting *Prunus* spp. fruit trees. *Journal of Fruit and Oriental Plant Research,* 15:117-124.

[14]   Clark, M. F. & Adams, A. N. (1977). Characteristics of the microplate method of enzyme-linked immunosorbent assay for the detection of plant viruses.*Journal of General Virology.,* 34: 475-483.

[15]   Damiano, C., Monticelli, S., Gentile, A., Di Nicola-Negri, E. & Ilardi, V. (2007). Efficiency of Prunus transformation for PPV-resistance by gene silencing and PPV coat protein gene strategies *Acta Horticulturae.,* 764: 63-70

[16]   Engel, G. (1990). The importance of virus-free plant material in integrated fruit production. *Acta Horticulturae,* 285: 127-133.

[17]   Eremin, G.V. & Podorozhnyi, V.N. (1992). Production of healthy planting material of cherry plum and stone fruit crop rootstocks. *Sadovodstvo-i-Vinogradarstvo.*8:11-12.

[18]   Faccioli, G. & Marani, F. (1998). Virus elimination by meristem tip culture and micrografting. . Eds. Hadidi, A., Khetarpal, R.K. and Koganezawa, H *In*: Plant virus disease control. USA. The American Phytopathological Society, 346-380.

[19] Fitch, M. M.M., Lehrer, A. Komor T.E., & Moore, P.H. (2001). Elimination of *Sugarcane yellow leaf curl virus* from infected sugarcane plants by meristem tip culture visualized by tissue blot immunoassay.*Plant Pathology.* 50: 676-680.

[20] Gabova, R. (1995). Chemotherapy treatment of *Prunus* spp. genotypes infected by *Plum pox virus. Rasteniev"dni-Nauki.*, 32:16-18.

[21] Ghosh, S.P. (2001). Temperate Fruit Production in India *Acta Horticulturae*, 565:131-138.

[22] Golino, D.A., Sim, S.T., Cunningham, M. & Rowhani, A. (2007). Transmission of rose mosaic viruses. *Acta Horticulturae.*, 751: 217-224.

[23] Hazaa, M.M., Dougdoug, Kh. A. El., & Maaty, Sabah Abo El. (2006). Eradication of banana viruses from naturally infected banana, production of certified banana plants and virus tested. *Journal of Applied Sciences Research,* 2: 714-722.

[24] Herranz, M. C., Sanchez-Navarro, J. A., Aparicio, F. & Pallás, V. (2005). Simultaneous detection of six stone fruit viruses by non-isotopic molecular hybridization using a unique riboprobe or 'polyprobe'. *Journal of Virological Methods*, 124: 49–55.

[25] Hilsendegen, P. (1999). Results of the tolerance of sour cherry cultivars to *Prunus necrotic ringspot virus* (PNRV). *Erwerbsobstbau.*, 41: 192-197.

[26] Kajati, I. & Szonyegi, S. (1996). The occurrence of and damage caused by *Prunus necrotic ringspot virus* (PNRSV) on sour cherries in Hungary in the 1960s and 1990s. *Acta Horticulturae.*, 410: 325-330.

[27] Kelley, R. D. & Cameron, H. R. (1986). Location of Prune dwarf and *Prunus necrotic ringspot viruses* associated with sweet cherry pollen and seed. *Phytopathology.* 76: 317-322.

[28] Knapp, E., Hanzer, V., Weiss, H., Camara-Machado, A.da, Weiss, Wang, Q., Katinger, H. & Laimer-da-Camara-Machado, M. (1995). New aspects of virus elimination in fruit trees.*Acta Horticulturae*, 386: 409-418.

[29] Kornova, K., Atanassova, S. & Todorov, A. (2002). Studies on the possibility of producing virus free planting material of plum cultivars by *in vitro* Propagation. *Acta Horticulturae.*, 577 : 187-193.

[30] Kunze, L. (1969). Der Einfluss der Stecklenberger Krankheit auf den Ertrag von *Sauerkirschen. Erverbsobstbau,* 11: 1-3.

[31] Laimer, M., Hanzer, V., Mendocna, D., Kristone, K., Toth, E.K., Kirilla, Z. & Balla, I. (2006).Elimination and detection of pathogens from tissue cultures of *Prunus* sp. *Acta-Horticulturae.*, 725 : 319-323.

[32] Lo´pez-Moya, Juan Jose´, Ferna´ndez-Ferna´ndez, Marı´a Rosario, Cambra, Mariano, Garcı´a & Juan Antonio. (2000). Biotechnological aspects of plum pox virus *Journal of Biotechnology.,* 76: 121–136

[33] Machado, M., Machado, A., Hanzer, V., Weiss, H., Regner, F., Steinkellner, H., Plail, R., Knapp, E. & Katinger, H. (1993). Coat Protein-Mediated Protection against *Plum pox virus. Acta Horticulturae.*, 336 : 85-92

[34] Mainardi, M. Gilli, G. & Triolo, E. (1992). Detection of *Prunus necrotic ring spot virus* and *Plum pox virus* in shoot tip cultures of Mr.S. 2/5 plum rootstock. *Advances-in-Horticultural-Science.*, 6: 173-175.

[35] Manganaris, G.A., Economou, A.S., Boubourakas, I.N. & Katis, N.I. (2003). Elimination of PPV and PNRSV through thermotherapy and meristem tip culture in nectarine *Cell Biology and Morphogenesis.*, 22: 195-200.

[36] Morel G. & Martin C. C. R. (1952). Virus-free Dahlia through meristem culture Hebd. Seances Academy of Science (Paris), 235: 1324-1325.

[37] Nacheva, L., Milusheva, S. & Ivanova, K. (2002). Elimination of *Plum pox virus* (PPV) in plum (*Prunus domestica* L.) cvs Kyustendilska sinya and Veljevka through *in vitro* techniques *Acta Horticulturae.*, 577 : 289-291.

[38] Németh, M. (1986). Virus, Mycoplasma and Rickettsia Diseases of Fruit Trees. Dordrecht, Boston, Lancaster: Martinus Nijhoff Publishers.pp.840

[39] Paunovic, S., Ruzic, D., Vujovic, T., Milenkovic, S. & Jevremovic, D. (2007). *In vitro* production of *Plum pox virus* free plums by chemotherapy with ribavirin *Biotechnol and Biotechnol EQ.*, 417-421.

[40] Pine, T.S. (1964). Influence of necrotic ringspot virus on growth and yield of peach trees. Phytopathology *Phytopathology,* 54: 504–505.

[41] Posnete, A.F. & Cropley, R. (1959). The reduction in yield caused by apple mosaic Annual Report East Malling Research Station, 89-90.

[42] Ramsdell, D. C., Bird. G. W., Adler, V. A. & Gillett, J. M. (1992). Effects of *Tomato ringspot virus* and *Prunus necrotic ringspot virus* alone and in combination on the growth and yield of 'Montmorency' sour cherry.*Acta Horticulturae,* 309: 111-114.

[43] Rana, T., Chandel, V., .Hallan, V. & Zaidi A.A. (2007a). Molecular evidence for *Apple Chlorotic Leaf Spot Virus in* wild and cultivated apricot in Himachal Pradesh, India. *Journal of Plant Pathology*, 89: 72.

[44] Rana, T.; Chandel, V.; Handa, A.; Thakur, P.D.; Hallan, V. & Zaidi A.A. (2007a). Molecular evidence of *Apple chlorotic leaf spot virus* in Himachal Pradesh. . *Indian Journal of Virology*, 18(2): 70-74

[45] Rana, T.; Chandel, V.; Hallan, V. & Zaidi A.A. (2007b). Molecular evidence for *Apple Chlorotic Leaf Spot Virus in* wild and cultivated apricot in Himachal Pradesh, India. *Journal of Plant Pathology*, 89: 72.

[46] Rana, T.; Chandel, V.; Hallan, V. & Zaidi A.A. (2008a). Characterization of *Apple chlorotic leaf spot virus* infecting almonds in India. *Australasian Plant Disease Notes*, 3: 65-67.

[47] Rana, T.; Chandel, V.; Hallan, V. & Zaidi A.A. (2008b). *Cydonia oblonga* as reservoir of *Apple chlorotic leaf spot virus* in India. *Plant Pathology* 156: 382-384.

[48] Rana, T.; Chandel, V.; Hallan, V. & Zaidi A.A. (2008c). Himalayan wild cherry (*Prunus cerasoides* D. Don) as a new host of *Apple chlorotic leaf spot virus. Forest Pathology*, 38 (2): 73-77.

[49] Rana, T.; Chandel, V.; Hallan, V. & Zaidi A.A. (2009) Molecular evidence for the presence of *Apple chlorotic leaf spot virus* in infected peach trees in India. *Scientia Horticulturae*, 120: 296-299.

[50] Ravelonandro, M., Scorza, R., Bachelier, J. C., Labonne, G., Levy, L., Damsteegt, V., Callahan, A. M., & Dunez, J. (1997). Resistance of transgenic *Prunus domestica* to *Plum pox virus* infection. *Plant Disease*, 81: 1231-1235.

[51] Reeves, E.L. & Cheney, P.W. (1962). Flowering cherries as symptomless hosts of little cherry virus. *In*: Proceedings of the Fifth European Symposium on Fruit Tree Virus Diseases, Bologna, Italy, pp. 108–114.

[52] Rizqi, A., Zemzami, M. & Spiegel, S. (2001). Recovery of virus free almond plants by improved *in vitro* shoot tip grafting *Acta Horticulturae.*, 550: 447-454.

[53] Rosner, A., Maslenin, L. & Spiegel, S. (1997). The use of short and long PCR products for improved detection of *Prunus necrotic ringspot virus* in woody plants. *Journal of Virological Methods.* 67: 135-141.

[54] Rosner, A., Maslenin, L. & Spiegel, S. (1998). Differentiation among *Prunus necrotic ringspot virus* by transcript conformation polymorphism. *Acta Horticulturae.*, 472 : 227-233

[55] Rowhani, A., Biardi, L. & Golino, D.A. (1998). Detection of viruses of woody host plants using colorimetric PCR. *Acta Horticulturae,* 472: 265-271.

[56] Salem, N., Mansour, A., Al-Musa, A. & Al-Nsour, A. (2003). Incidence of *Prunus necrotic ringspot virus* in Jordan. *Phytopathologia Mediterranea.*, 42: 275-279.

[57] Scorza, R, Revalonandaro, M, Callahan, A.M., Cordts, J.M., Fuchs, M., Dunez, J. & Gonsalves, D. (1994). Transgenic plums (*Prunus domestica*) express the *Plum pox virus* coat protein gene. *Plant Cell Reports*, 14: 18-22.

[58] Sharma, S., Singh, B., Rani, G., Zaidi, A.A., Hallan, V., Nagpal, A. & Virk, G.S. (2008). In vitro production of *Indian citrus ringspot virus* (ICRSV) free Kinnow plants employing thermotherapy coupled with shoot tip grafting . *Plant Cell Tissue and Organ Culture.*, 92: 85-92.

[59] Spiegel, S., Scott, S. W., Bowman, V. V., Tam, Y., Galiakparov, N. N. & Rosner A. (1996). Improved detection of *Prunus necrotic ringspot virus* by the polymerase chain reaction.*European Journal of Plant Pathology,* 102: 681-685.

[60] Stouffer, R.F. & Smith, S.H. (1971). Present status of the Prunus stem pitting diseases in the United States. Proceedings of the European Symposium on Fruit Tree Virus Disease, pp. 109–116.

[61] Sutic, D. D., Ford, R.E. & Tosic, M.T. (1999). Handbook of Plant Virus Diseases. United States of America. CRC Press LLC. 553pp.

[62] Tenllado F, Llave C, Ramon J. & Ruiz D. (2004). RNA interference as a new biotechnological tool for the control of virus diseases in plants. *Virus Research.*, 102: 85–96.

[63] Thomas, B. J. (1980). The detection by serological methods of viruses infecting rose. *Annals of Applied Biology,* 94: 91-101.

[64] Topchiiska, M. (1983). Effect of *Prunus necrotic ringspot virus* (PNRSV) and *Prune dwarf virus* (PDV) on some biological properties of peach. *Acta Horticulturae,* 130: 307-312.

[65] Verma, N., Ram, R. & Zaidi, AA. (2005). *In vitro* production of *Prunus necrotic ring spot virus* free begonias through chemo and thermotherapy. *Scientia Horticulture Scientia Horticulture*, 103: 239-247.

[66] Verma, N. Ram, R.Hallan, V., Kumar, K. & Zaidi, A.A. (2007). Production of *Cucumber mosaic virus*-free chrysanthemums by meristem tip culture. *Crop Protection.*, 23 : 469-473.

[67] Vertesy, J. (1980). *In vitro* propagation of *Prunus persica* and *P. persico-davidiana* shoot tips in order to get virus-free plants. *Acta-Phytopathologica-Academiae-Scientiarum-Hungaricae.*, 15: 261-264

[68] Wang, J. X, Liu, Z., Xie, X. H., Wu, B. & Tong, Z.G. (2000). The reaction of virus-free apple tree growth and fruit production. *Acta-Horticulturae-Sinica.*, 27: 157-160

[69] Wei, S. & Timon, B. (1994). Improvement of efficiency in producing virus-free peach (*Prunus persica*) plants *in vitro.Jiangsu-Journal-of-Agricultural-Sciences*, 10: 1-4.

[70] Wood, G. A. & Fry, P. R. (1984). Effect of virus infection on growth and cropping of Greengage and Billington plum trees. *New Zealand Journal of Agricultural Research,* 27: 563-568.

[71] Yamaga, H. & Munakata, T. (1991). Production of virus-free apple planting stock by meristem culture. *Technical-Bulletin Food-and-Fertilizer-Technology-Center.*, 126: 10-17.

[72] Zotto, A. dal & Docampo, D.M. (1997). Micropropagation of rootstocks of plum cv. Marianna 2624 (*Prunus cerasifera x P. munsoniana*) and Pixy (*P. insititia* L.) of certified sanitary status. *Phyton-Buenos-Aires.*, 60 : 127-135.

In: Recent Trends in Biotechnology and Microbiology
Editors: Rajarshi Kumar Gaur et al., pp. 81-86

ISBN: 978-1-60876-666-6
2010 Nova Science Publishers, Inc.

Chapter 7

TRANSGENIC CROPS- ANOTHER LEAPFROG TECHNOLOGY

P. Balasubramanian

Professor and Head, Department of Plant Molecular Biology and Biotechnology
Centre for Plant Molecular Biology, Tamil Nadu Agricultural University
Coimbatore 641 003

ABSTRACT

The possibility to transfer genes across almost all taxonomic borders by molecular techniques has expanded the potential resources available to plant breeders enormously. It is becoming generally accepted that a multidisciplinary approach to plant biology will lead to the disappearance of borders between disciplines. In the same vein, the differences between transgenic and non-transgenic crops should become irrelevant when the focus of plant breeding is on achieving maximal production in a sustainable way to feed the growing human population.

Dubbed 'the Green Phoenix', the transgenic plant technology offers both challenges and opportunities for growth and development of mankind. This technology should be used to complement the traditional methods for enhancing productivity and quality, rather than to replace the conventional methods. To adopt this technology, GM crops and their products, awareness has to be created among the farming and consumer communities regarding their benefits and effects on human life, by the scientific communities and national leaders. This presentation also includes a few recent advancements in transgenic plant technology and their potential role in enhancing the quality of human life.

INTRODUCTION

Ring farewell to the century of physics, the one in which we split the atom and turned silicon into computing power. It's time to ring in the century of biotechnology. Just as the discovery of the electron in 1897 was a seminal event for the 20th century, the seeds for the 21st century were spawned in 1953, when James Watson blurted out to Francis Crick how four nucleic acids could pair to form the self-copying code of a DNA molecule. Now we have accomplished one of the most important breakthroughs of all time: deciphering the human genome, the 100,000 genes encoded by 3 billion chemical pairs in our DNA. This is the era of biotechnology and it has made several gigantic leaps since its beginning in 1953 to make the greatest achievements like human genome sequencing.

OVERVIEW OF BIOTECHNOLOGY

The term biotechnology covers a wide range of scientific techniques and products that can be used in numerous ways to boost and sustain the productivity of crops, livestock, fisheries and forests. It is the technique of using living organisms or their parts to make or modify products, improve plants or animals or develop micro- organisms for specific use. The practice of biotechnology in its varied forms, varying in style, substance and scale, reveals shared histories with the planet's peoples and cultures in the developed and developing worlds. Biotechnology has a long history. Old as the growing of crops and the making of cheese and the production of wines, its practice has been described as one of the oldest professions in the world. Modern Biotechnology, in general, embraces more often the principles and practice of genetic engineering. Several other terms such as genetic engineering; genetic transformation; transgenic technology, recombinant DNA technology, and genetic modification (GM) technology have been used to describe the applications of this form of modern biotechnology.

The modern biotechnology broadly has two groups of technologies. The cellular approaches include tissue culture, which is used in micro propagation and animal reproduction. The molecular approaches include genomics and bioinformatics, diagnostic procedures, molecular markers technology and genetic engineering. Other biotechnologies include use of bio-fertilizers and bio pesticides. Genetic modification (GM) refers to approaches within the broad domain of biotechnology, which are distinguished from other biotechnology techniques by allowing the transfer of genes between different organisms

GENETICALLY ENGINEERED ORGANISMS: THE NEXT GREEN REVOLUTION? TRANSGENIC PLANTS

The development of gene transfer technology allowed plant scientists to intervene in a precise manner to genetically modify individual embryo cells that could then become the parents of countless other plants, each of which would also contain the newly inserted gene. Among the first important advance was the discovery of a bacterium called *Agrobacterium tumifaciens*, which has the natural capacity to penetrate the cell wall, was an excellent vehicle with which to transfer a desirable gene from another species. In 1970, the restriction enzymes were discovered which can cut DNA at precise spot which made the *Agrobacterium* mediated genetic engineering easy. In 1994 scientists succeeded in using *Agrobacteriumt* to create transgenic rice. Since then scientists have succeeded in using similar approach to create a variety of transgenic plants. In the last few years, scientists have also found other ways to transfer genes like electroporation, particle bombardment and *in planta* transformation techniques.

Botanists began creating transgenic plants with Bt genes in the mid 1980s and by 1990s they had shown that the proteins coded by the gene protect tobacco, tomato and potato plants from certain pests. Since then, they have made remarkable progress by redesigning the Bt genes to produce much larger amount of insecticidal protein. Most of the transgenic technologies for plants are for disease and pest resistance.

Insect-Resistant Bt Cotton Leads the way

Transgenic cotton varieties containing insect-resistance genes derived from the insecticidal microbe *Bacillus thuringiensis* are now being grown commercially in China, South Africa, Mexico, Argentina, Indonesia, and India. Pray *et al.* The rapid spread of Bt cotton was driven by the farmers' demand for a technology that increases yield, reduces insecticide use and costs, reduces insecticide poisonings and requires less labor. Initial yield increases were in the 5–10% range and modest increases continued over time, suggesting that farmers are learning to manage Bt varieties better. There is no indication that insect pests are becoming resistant to Bt cotton

Disease Resistance

Several field tests of transgenic crops containing genes for bacterial and fungal disease resistance are under way in developing countries and results have been promising so far. However, none of these crops has yet been commercialized. Rather, the first biotechnology-derived disease resistant lines to be commercialized have resulted from pyramiding natural resistance genes via marker-assisted selection (MAS).In January 2002, the government of Indonesia released two new rice varieties, 'Angke' and 'Conde', which were derived by disease resistance breeding augmented with polymerase chain reaction (PCR)-based MAS to pyramid bacterial blight resistance genes into commercially adapted varieties. These new varieties are the product of more than ten years of international collaborative research efforts that led to a better understanding of population genetics and genome structure of the rice pathogen 'bacterial blight' (*Xanthomonas oryzae*) and to an increasing inventory of bacterial blight resistance genes from the genomes of *Oryza sativa* L. and near relatives such as *Oryza minuta*

PRESENT STATUS OF TRANSGENIC CROPS

The first transgenic plant product marketed commercially was the well-known 'FlavrSavr' tomato which had been modified to contain reduced levels of the cell wall softening enzyme polygalacturonase. Tomato with a similar type of modification has been on the market in the UK since February 1996. Since that time, however, there has been a massive expansion in the growth of transgenic field crops, particularly maize, soybean, oilseed rape, and cotton, such that in North America transgenic varieties now represent the majority of the acreage of these crops. For example, it is now estimated that 70% of the Canadian oilseed rape crop in 1999 will be genetically modified. Most of the varieties grown to date have been modified with genes for herbicide or insect resistance, both of which provide a significant economic benefit for the farmer. Material from these modified crops, particularly soybean and maize, is now imported into the European Union in large quantities. However, in several European countries there is now considerable consumer resistance to food products containing such GM constituents, either in the native form or as processed derivatives (e.g. soya lecithin). In the UK this resistance seems to be based primarily on a mistrust of the government regulatory process, allied to a strong desire for choice (to date the commodity GM products have not been segregated from their non-GM equivalent). These concerns have led to many retailers removing GM soya and maize products from sale, though enzymes and vitamins from GM microbes remain as common components of many foods and drinks. The other aspects, which have been addressed with transgenic approach, include the following photosynthetic enhancement, chloroplast transformation, sugar and starch metabolism and alteration in senescence.

Plant Factories

The emergence of transgenic technology raised the fascinating possibility that plants could be designed as factories to produce metabolites at low cost. In 1996 the research on this aspects started, a biotech company called crop tech produced human glucocerebrosidase in tobacco. There after several researches has done on this aspect which produced human heme factors, edible vaccines and plastics, dyes and biofuels in plants. Much attention has been given recently to the potential of using plants as a production system for high value compounds of medical importance. Examples include the expression of antibodies that may reduce the growth of bacteria associated with dental caries or that can be used in the treatment of human tumours. Other medically related projects include the production of recombinant blood factors, human milk proteins, interferon (Ohya et al., 2001) gonadotrophin (Abdennebi-Najar et al., 2001) and the human secretory protein somatotrophin (Moloney and Habibi, 2001)

Biodegradable Plastics

One of the recurring themes concerning GM crops is discussion of their potential as a production system for biodegradable plastics (Poirier, 2001), particularly poly (beta-hydroxybutyrate) (PHB) and poly (beta- hydroxyvalerate) (PHV). These projects, which were initially developed as alternatives to an expensive, bacterial fermentation system, have progressed to the extent that PHB concentrations as high as 7.7% of fresh seed weight have been reported in oilseed rape. Such levels were achieved by coordinated expression of the three bacterial enzymes, beta-ketothiolase, acetoacyl- CoA reductase and PHB synthase in the leucoplasts of mature seeds.

FUTURE COMMERCIAL TRENDS- NUTRITIONAL GENOMICS AND TREE BREEDING

These are the next two challenges which should be accomplished in the near future of plant biotechnology. This includes biochemical pathway engineering and increasing the dietary supplements in basic foods rice and wheat. Golden rice project, folate fortification all these come under this aspect. Despite progress with supplementation and fortification programs, there is compelling evidence that persistent deficiencies of iron, zinc, iodine and vitamins remain a major cause of numerous human health problems in developing countries. For example, a recent analysis indicates that 127 million pre-school children still suffer from vitamin A deficiencies, leading to blindness and early death. Now, through advances in plant biotechnology there are new opportunities to complement supplementation by including enhanced human nutrition along with higher yields, reduced losses and greater tolerance of adverse growing conditions as an important objective when developing crop varieties.

Beyer and coworkers from Germany reported further advances in the development of 'Golden Rice': transgenic lines that are engineered to synthesize provitamin A (β-carotene) in the rice endosperm.β-carotene synthesis was achieved by adding only two genes, daffodil phytoene synthase (*psy*) and bacterial phytoene desaturase (*crtI*), with endosperm-specific promoters. These new 'clean' lines are being sent to collaborating breeding programs in Asia where they will be crossed with local varieties that are well adapted to the regions where vitamin A deficiency is still prevalent

CONCLUSIONS

From a human welfare standpoint, the greatest benefits of plant biotechnology will surely be derived from the adoption of improved crop varieties in the developing countries of the world where billions of people still depend on agriculture for their livelihoods. Transgenic have a key role in curing some life threatening diseases which in other way is very difficult to cure. The genetic revolution will be remembered as one of the great ascents of the human mind. This is a journey, which is just started and ultimately will lead us to a world in which we will be able to influence our own evolution.

REFERENCES

[1] Abdennebi-Najar, L., Bakker, H.A., Bosch, H., Dirnberger, D., Remy, J.J., Steinkellner, H. & Van De Wiel, D.F.M. (2001). *Gonadotrophins in plants.* Patent Application WO 01/31044.

[2] Dunwell J.M. (1999). Transgenic approaches to crop improvement. *Journal of experimental Botany*, 51: 487-496.

[3] Moloney, M.M. & Habibi, H.R. (2001). Expression of somatotrophin in plant seeds. US Patent 6:288-304.

[4] Poirier, Y. (2001). Production of polyesters in transgenic plants. *Advances in Biochemical Engineering*. 71: 209–240.

[5]   Ohya, K., Matsumura, T., Ohashi, K., Onuma, M. & Sugimoto, C. (2001). Expression of two subtypes of human IFN-alpha in transgenic potato plants. *J. Interferon Cytokine Res.* 21: *595– 602.*

[6]   Reilly, P. R. (2000). Abraham Lincoln's DNA and other adventures in genetics, Cold Spring Harbor Laboratory Press.

[7]   Toenniessen, G. H, Johns, C.O. & DeVries, J. (2003). Advances in plant biotechnology and its adoption in developing countries, *Current opinions in biotechnology,* 6:191– 198.

[8]   Uzogara, S.G. (2000). The impact of genetic modification of human foods in the 21st century: A review. *Biotechnology Advances* 3: 179-206.

In: Recent Trends in Biotechnology and Microbiology

Editors: Rajarshi Kumar Gaur et al., pp. 87-108

ISBN: 978-1-60876-666-6

2010 Nova Science Publishers, Inc.

Chapter 8

PHYTOPLASMA DISEASES OF WEEDS: DETECTION, TAXONOMY AND DIVERSITY

Smriti Mall[*1], *Govind P. Rao*[1] *and Carmine Marcone*[2]

Sugarcane Research Station, Kunraghat, Gorakhpur 273 008, UP, India[1]

Dipartimento di Scienze Farmaceutiche, Università degli Studi di Salerno, Via Ponte Don Melillo, I-84084 Fisciano (Salerno), Italy[2]

ABSTRACT

Phytoplasmas cause diseases in several weeds which may act as alternative natural hosts facilitating the spread of phytoplasmas to other economically important plants and thereby increasing economic losses. The most peculiar symptoms observed on weeds include extensive chlorosis, proliferation of axillary shoots, witches'-broom, yellowing and little leaves. So far more than 43 weed species were reported having phytoplasma infections from all over the world. Nucleotide sequence studies have shown that weed-infecting phytoplasmas mainly belongs to 16SrI, 16SrIV, 16SrVIII, 16SrXI and 16SrXIV groups. Among them, 16SrI and 16SrXIV phytoplasmas have a more wide occurrence in nature all over the world. Even though the weeds identified as phytoplasma hosts often grow abundantly around field crops, the possibilities of transmission of phytoplasmas related to important agricultural, economical and horticultural crops from weed to the mentioned crops and vice-versa can not be ignored. This could be because phytoplasmas are able to survive in many potential economical crops or because an insect vector is capable of transmitting phytoplasmas from other weeds to crops which are already known as phytoplasma hosts. In either case, the chance of transmission in the future seems high, given the large phytoplasma reservoir already revealed, the propensity of new phytoplasma strains to evolve. In this chapter, detailed and up-to-date information on occurrence, symtomatology, molecular characterization, transmission, taxonomy, genetic diversity and management approaches on weed phytoplasmas has been discussed. Knowledge of the diversity of phytoplasmas will be expanded by recent studies and the availability of molecular tools for pathogen identification.

[*] Corresponding author. E-mail: gprao_gor@rediffmail.com

Introduction

Plant-pathogenic phytoplasmas are wall-less, unculturable bacteria of the class *Mollicutes* that are associated with over 1,000 plant diseases worldwide (Seemuller *et al.*, 1998; Al-Saady and Khan, 2006; Lee *et al.*, 2007; Harrison *et al.*, 2008). They have a small genome that ranges in size from 530 to 1350 kilobases, a G+C content between 24 and 33 mol%, two rRNA operons, a low number of tRNAs and a limited set of metabolic enzymes (Bove, 1997; Marcone *et al.*, 1999;; Oshima *et al.*, 2004; Bai *et al.*, 2006). Phytoplasmas are obligate parasites which could survive and multiply only in their plant hosts and insect vectors. According to phylogenetic studies, phytoplasmas are more closely related to *Acholeplasmas* than to other mollicutes. *Acholeplasma palmae* and *A. modicum* are the closest known relatives of phytoplasmas. Diseases induced by phytoplasmas are suggestive of profound disturbances in the normal balance of growth regulators, are characterized by flower malformation, growth aberrations, yellowing and/or decline, and are collectively referred to as yellows disease. These diseases were thought to be caused by viruses until a group of Japanese scientists in 1967 discovered that they were associated with pleomorphic bodies, ranging in size from 200 to 800 nm, which were then termed mycoplasma-like organisms (Doi *et al.*, 1967).

The phytoplasma detection was relied for more than two decades on DAPI staining and electron microscopy. However, in the last 20 years the applications of DNA-based technology allowed to distinguish different phylogenetic groups inside the phytoplasma clade. The Phytoplasma Working Team of the International Research Project for Comparative Mycoplasmology (IRPCM) adopted the trival name 'phytoplasma' to replace the term mycoplasma-like organisms. Later on, the 'Candidatus Phytoplasma' genus has been proposed and adopted in order to start formal classification of these prokaryotes, some of them being associated with important or quarantine-subjected plant diseases (IRPCM, 2004). Up-to-date satisfaction of Koch's postulates has not been achieved, but several indirect evidences confirmed that they are responsible many plant diseases worldwide. It was also demonstrated that genetically undistinguishable phytoplasmas can be associated with diseases inducing different symptoms and/or affecting different plant species whereas genetically different phytoplasmas can be associated with similar symptoms in the same or in different plant host(s).

Phytoplasma DISEASES of Weeds

Several weeds are reservoirs of important phytoplasmas causing serious diseases to important commercial crops and play an important role in spreading phytoplasmas and serve as natural alternative hosts (Blanche *et al.*, 2003; Joomun *et al.*, 2007; Pasquini *et al.*, 2007; Harrison *et al.*, 2008). Phytoplasmas cause diseases in several weeds, which have resulted as alternative natural host facilitating the spread of phytoplasmas to other economically important plants and thereby increasing economic losses. Early detection of these phytoplasmas associated with diseases of weed crops is very important to check the possibility of further spread of phytoplasma diseases to other commercial crops.

Many weeds infected by phytoplasmas are known to show a wide range of symptoms such as white leaf, yellows, witches'-broom and phyllody (Dabek, 1983; Shiomi *et al.,* 1983; Choi *et al.,* 1985; Reddy and Jeyarajan, 1990; Gibb *et al.,* 1997; Tran-Nguyen *et al.,* 2000; Arocha *et al.,* 2005; Marcone and Rao, 2008). Among the several weed diseases, Bermuda grass white leaf (BGWL) is a destructive phytoplasmal disease of Bermuda grass, *Cynodon dactylon* (L.) Pers., which was first reported from Taiwan (Chen *et al.,* 1972). The disease is known to occur in several Asian countries (Chen *et al.,* 1972; Zahoor *et al.,* 1995; Sdoodee *et al.,* 1999; Davis and Dally, 2001; Jung *et al.,* 2003; Rao *et al.,* 2007), Sudan (Dafalla and Cousin, 1988; Cronje *et al.,* 2000a, b), Italy (Marcone *et al.,* 1997, 2004), Cuba (Arocha *et al.,* 2005) and Australia (Schneider *et al.,* 1999; Tran-Nguyen *et al.,* 2000; Blanche *et al.,* 2003). It was also known as *Cynodon* white leaf (CWL). The causal agent, the BGWL phytoplasma, is a member of the phylogenetic BGWL phytoplasma group or 16SrXIV group, which includes several other poaceaous plant pathogens, such as brachiaria grass (*Brachiaria distachya*) white leaf (BraWL), annual blue grass *(Poa annua)* white leaf (ABGWL), *Dactyloctenium (Dactyloctenium aegyptium)* white leaf (DacWL) and carpet grass *(Axonopus compressus)* white leaf (CGWL) phytoplasmas, tassel blue grass (*Dichanthium annulatum*) white leaf (Lee *et al.,* 1997, 1998, 2000; Seemuller *et al.,* 1998; Sdoodee *et al.,* 1999; Marcone *et al.,* 2004; Rao *et al.,* 2009). The BGWL phytoplasma is a discrete taxon at the putative species level, for which the name 'Candidatus Phytoplasma cynodontis' has been proposed (Marcone *et al.,* 2004; Arocha *et al.,* 2005.

Arocha *et al.* (2005) first reported on the occurrence of phytoplasmas in Bermuda grass in Cuba and also observed that BGWL disease there is caused by a phytoplasma belonging to the 16SrXIV group. Marcone *et al.* (1997) detected BGWL disease in Italy and characterized the association   phytoplasma by RFLP analysis. Sarindu and Clark (1993) reported the occurrence of sugarcane white leaf and BGWL diseases from Thailand. Gibb *et al.* (1997) detected and identified phytoplasmas associated with white leaf and stunting disease in Australian grasses. Livingston *et al.* (2006) first reported a 16SrII group phytoplasma in *Polygala mascatense*, a weed in Oman. Some *P. mascatense* plants showed stunting, small leaves, bushy growth and phyllody symptoms. PCR assay confirmed the presence of phytoplasma infections causing witches'-broom in *Polygala*. RFLP results showed that the *Polygala*-infecting phytoplasma is similar to Lucerne phytoplasma (16SrII) group. Phytoplasmas were observed in phloem of naturally infected plants of the weed *E. cicutarium* collected in the semi-arid coastal area of Chile. The study revealed numerous straight and curved portions of filamentous phytoplasma bodies, but no helical bodies were observed (Graf *et al.,* 1978). Lee *et al.* (1992) reported on phytoplasma infecting *Trifolium* sp. (Fabaceae) in Canada with the typical clover phyllody symptoms and identified the causal agent as a 16SrI-C aster yellow group member.

Smrze *et al.* (1981) found   that   phytoplasmas sporadically occur i lucerne in Czechoslovakia. The disease was graft-transmitted to healthy plants and the symptoms, in comparison to those reported from Australia and USA, were mild and inconspicuous. Phytoplasma bodies were found in diseased but not in healthy phloem tissues. Arocha *et al.* (2006) reported phytoplasma infections in *Ocimum basilicum* (Labiatae) growing wildly in Cuba showing little leaf symptoms.

Brown *et al.* (2008) reported a phytoplasma which is a member of the 16SrIV group in *Emelia fosbergii, Synedrella nodiflora* and *Vernonia cineria* (Asteraceae) in Jamaica, showing lethal yellowing symptoms. Reeder *et al.* (2008) reported a phytoplasma in yellows-diseased *Senecio jacobaea* in UK which belongs to 16SrI group. Koh *et al.* (2008) found occurrence of phytoplasma on *Paspalum conjugatum* (Poaceae) in Singapore showing white leaf disease symptoms.

Singh *et al.* (1978) revealed phytoplasma bodies in phloem cells of diseased *C. dactylon* plants collected near Varanasi, U.P., India and proved the causative role of the phytoplasma bodies with the disease. Reddy and Jeyarajan (1990) found rice yellow dwarf symptoms caused by phytoplasma in the weed, *Echinochloa colonum*. They also successfully transmitted the phytoplasma from rice to this weed.

Muniyappa *et al.* (1979,1982) observed phytoplasma in *C. dactylon* and yellowing disease of *Urocha panicoides* in South India . Pleomorphic phytoplasma bodies were present in phloem tissue of infected plants. They found that application of oxytetracycline hydrochloride suppressed the symptoms temporarily. Padmanabhan (1982) reported a phyllody disease of *Parthenium hysterophorus* from South India. The disease is characterized by witches'-broom symptoms. Transmission of the associated phytoplasma from diseased to healthy plants using the insect vector *Hishimonus phycitis* was also accomplished. Rao *et al.* (1990) reported on a white leaf disease of *Imperata arundinacea* caused by phytoplasma. *Cannabis sativa* (Cannabinaceae) and *Achyranthes aspera* (Amaranthaceae) showing witches'-broom and yellowing symptoms in India were also found to be associated with 16SrI group phytoplasmal infections (Raj *et al.*,2008; Mall, 2009). Other reported weed diseases from India are leaf phyllody of *Phyllanthus fraternus,* grassy and bunchy shoot of *Cenchrus ciliaris,* bunchy shoot of *Dactyloctenium aegyptium* and white leaf of *Imperata arundianacea* (Mall *et al.,* 2007).

A witches'-broom disease caused by phytoplasma on *Bupleurum falcatum, Cnidium officinale* and *Plantago asiatica* has been observed in Korea (Choi *et al.,* 1985) whereas a little leaf disease of *Rhynchosia minima* associated with phytoplasma has been reported fom Jamaica (Dabek, 1983).

GEOGRAPHICAL DISTRIBUTION

Phytoplasmas cause diseases in several weeds, which act as alternative host for the spread of phytoplasmas to other economically important plants and thereby possibilities of causing severe losses. So far, 43 phytoplasmas belonging to 17 groups have been identified in weed species (Table 1 and 2). An up-to-date information on different phytoplasmas reported on weeds with respect to their distribution are listed in (Table 1).

Table 1. Worldwide distribution of phytoplasmas in weeds

S. No.		Plant host	Disease	Geographic origin	Reference
1.	24.	*Acanthospermum hispidium* (Asteraceae)	Little leaf	India	Raju and Muniyappa (1981)
2.	28.	*Achyranthes aspera* (Amaranthaceae)	Yellows disease	India	Raj *et al.* (2008)
3.	4.	*Asclepias* sp. (Asclepiadaceae)	Milkweed yellows (MWY)	New York	Griffiths *et al.* (1994) ; Gunderseon *et al.* (1994)

4.	4.	*Asclepias* sp. (Asclepiadaceae)	Milkweed yellows (MWY)	New York	Griffiths *et al.* (1994) ; Gunderseon *et al.* (1994)
5.	14.	*Axonopus compressus* (Poaceae)	Carpet grass white leaf (CGWL)	Thailand Singapore	Sdoodee *et al.* (1999); Koh *et al.* (2008)
6.	13.	*Brachiaria distachya* (Poaceae)	Brachiaria white leaf (BraWL)	China	Chen *et al.* (1972)
7.	15.	*Bupleurum falcatum* (Apiaceae)	Bupleurum witches'-broom	Japan	Shiomi *et al.* (1983)
8.	26.	*Cannabis sativa* (Cannabinaceae)	Witches'-broom	India	Raj *et al.* (2008); Mall *et al.* (2009)
9.	34.	*Cassia italica* (Fabaceae)	Witches'-broom	Oman	Khan *et al.* (2007)
10.	36.	*Cenchrus ciliaris* (Poaceae)	Grassy and bunchy shoot	India	Mall *et al* . (2007)
11.	39.	*Chloris inflate* (Poaceae)	Creamy leaf	Australia	Blanche *et al* . (2003)
12.	29.	*Crotalaria tetragona* (Fabaceae)	Witches'-broom	India Brazil	Baiswar *et al.* (2009); Ribeiro *et al.* (2001)
13.	7.	*Cynodon dactylon* (Poaceae)	Bermuda grass white leaf (BGWL)	Thailand Cuba Italy Sudan Australia Singapore India	Lee *et al.* (1997) Arocha *et al.* (2005) Marcone *et al.* (1997) Dafalla and Cousin (1988) Schneider *et al.* (1999) Koh *et al.* (2008) Rao *et al* . (2007)
14.	37.	*Dactyloctenium aegyptium* (Poaceae)	Bunchy shoot	India	Mall *et al* . (2007)
15.	40.	*Dactyloctenium aegyptium* (Poaceae)	Grassy shoot disease	Australia	Blanche *et al* . (2003)
16.	31.	*Dichanthium annulatum* (Poaceae)	White leaf	India	Rao *et al.* (2009)
17.	16.	*Echinochloa colonum* (Poaceae)	Rice yellow dwarf (RYD)	India	Reddy and Jeyarajan (1990)
18.	9.	*Emelia fosbergii* (Asteraceae)	Lethal yellowing	Jamaica	Brown *et al.* (2008)
19.	32.	*Erigeron bonariensis* (Asteraceae)	Witches'-broom	Brazil	Davis *et al.* (1994)
20.	23.	*Erodium cicutarium* (Geraniaceae)	Yellow disease	Chile	Graf *et al.* (1978)
21.	38.	*Imperata arundianacea* (Poaceae)	White leaf	India	Mall *et al* . (2007)
22.	17.	*Imperata arundinacea* (Poaceae)	White leaf disease	India	Rao *et al.* (1990)
23.	6.	*Medicago sativa* (Fabaceae)	Loofah witches'-broom (LfWB)	Taiwan	Lee *et al.* (1993a)

Table 1. (Comtinued)

24.	21.	*Ocimum* sp. (Labiatae)	Basil little leaf	Cuba	Arocha *et al.* (2006)
25.	18.	*Parthenium hysterophorus* (Asteraceae)	Phyllody disease Witches'-broom	India India	Padmanabhan (1982) Raj *et al* . (2008)
26.	25.	*Paspalum conjugatum* (Poaceae)	White leaf disease	Singapore	Koh *et al.* (2008)
27.	20.	*Pennisetum purpureum* (Poaceae)	Napier grass stunt disease	Uganda	Nielsen *et al.* (2007)
28.	35.	*Phyllanthus fraternus* (Euphorbiaceae)	Leaf phyllody	India	Mall *et al* . (2007)
29.	8.	*Poa annua* (Poaceae)	Annual blue grass white leaf (ABGWL)	Italy	Lee *et al.* (1997)
30.	12.	*Polygala mascatense* (Polygalaceae)	Polygala witches'-broom	Oman	Livingston *et al.* (2006)
31.	22.	*Rhynchosia minima* (Fabaceae)	Rhynchosia little leaf	Jamaica	Dabek *et al.* (1983)
32.	30.	*Sasa fortunei* (Poaceae)	Witches'-broom	China	Zhang *et al.* (2009)
33.	11.	*Senecio jacobaea* (Asteraceae)	Aster Ragwort yellows	UK	Reeder *et al.* (2008)
34.	33.	*Sicana odorifera* (Cucurbitaceae)	Witches'-broom	Brazil	Montano *et al.* (2000)
35.	41.	*Sorghum stipoidem* (Poaceae)	White leaf	Australia	Blanche *et al* . (2003)
36.	10.	*Synedrella nodiflora* (Asteraceae)	Lethal yellowing	Jamaica	Brown *et al.* (2008)
37.	1.	*Trifolium* sp. (Fabaceae)	Clover Phyllody (Cph)	Canada	Lee *et al.* (1992)
38.	3.	*Trifolium* sp. (Fabaceae)	Clover yellow edge (CYE)	Canada	Lee *et al.* (1992, 1993)
39.	1.	*Trifolium* sp. (Fabaceae)	Clover Phyllody (Cph)	Canada	Lee *et al.* (1992)
40.	3.	*Trifolium* sp. (Fabaceae)	Clover yellow edge (CYE)	Canada	Lee *et al.* (1992, 1993)
41.	5.	*Trifolium* sp. (Fabaceae)	Clover proliferation (CP)	Canada	Deng *et al.* (1991) ; Lee *et al.* (1991)
42.	19.	*Urochloa panicoides* (Poaceae)	Yellowing disease	India	Muniyappa *et al.* (1982)
43.	27.	*Vernonia cineria* (Asteraceae)	Lethal yellowing	Jamaica	Brown *et al.* (2008)
44.	2.	*Vicia faba* (Fabaceae)	Faba been Phyllody(FBP)	Sudan	Schneider *et al.* (1995)
45.	2.	*Vicia faba* (Fabaceae)	Faba been Phyllody(FBP)	Sudan	Schneider *et al.* (1995)
46.	43.	*Whiteochloa biciliata* (Poaceae)	White leaf	Australia	Blanche *et al.* (2003)
47.	42.	*Whiteochloa cymbiforms* (Poaceae)	White leaf	Australia	Blanche *et al* . (2003)

DETECTION OF PHYTOPLASMAS

Symptoms

Weeds associated with phytoplasma infections show a great variety of symptoms. Most characteristic symptoms are extensive chlorosis, proliferation of axillary shoots, bushy growing habit, little leaves, shortened stolons and rhizomes, generalized stunting, virescence/phyllody, witches'-broom and death of the entire plants. In the early stage of the disease, light green to yellow streaks on the leaves are also present in some weeds associated with phytoplasma disease (Singh *et al.*, 1978; Muniyappa *et al.*, 1979; Sarindu and Clark, 1993; Marcone *et al.*, 1997; Gibb *et al.*, 1997;; Arocha *et al.*, 2005; Rao *et al.*, 2007, 2009; Mall *et al.*, 2009). A different range of symptoms produced by phytoplasmas on different weeds all around the world are listed in Table 2. Some common phytoplasma symptoms are also depicted in Fig. 1.

Table 2. Symptoms of phytoplasmas occurring in weeds.

S. No.		Weed Species	Disease	Symptoms	Reference
48.	24.	*Acanthospermum hispidium* (Asteraceae)	Little leaf	Little leaf	Raju and Muniyappa (1981)
49.	28.	*Achyranthes aspera* (Amaranthaceae)	Yellowing disease	Yellowing	Raj *et al.* (2008)
50.	4.	*Asclepias* sp. (Asclepiadaceae)	Milkweed yellows (MWY)	Yellowing	Griffiths *et al.* (1994); Gunderseon *et al.* (1994)
51.	14.	*Axonopus compressus* (Poaceae)	Carpet grass white leaf (CGWL)	White leaf	Sdoodee *et al.* (1999)
52.	13.	*Brachiaria distachya* (Poaceae)	Brachiaria white leaf (BraWL)	White leaf	Chen *et al.* (1972)
53.	15.	*Bupleurum falcatum* (Apiaceae)	Bupleurum witches'-broom	Yellowing and Witches'-broom	Shiomi *et al.* (1983)
54.	26.	*Cannabis sativa* (Cannabinaceae)	Wwitches'-broom disease	Wwitches'-broom	Raj *et al.* (2008)
55.	34.	*Cassia italica* (Fabaceae)	Witches'-broom disease	Witches'-broom	Khan *et al.* (2007)
56.	36.	*Cenchrus ciliaris* (Poaceae)	Grassy and bunchy shoot disease	Grassy and bunchy shoot	Mall *et al.* (2007)
57.	39.	*Chloris inflate* (Poaceae)	Creamy leaf disease	Creamy leaf	Blanche *et al.* (2003)
58.	29.	*Crotalaria tetragona* (Fabaceae)	Wwitches'-broom disease	Wwitches'-broom	Baiswar *et al.*, (2009); Ribeiro *et al* . (2001)
59.	7.	*Cynodon dactylon* (Poaceae)	Bermuda grass white leaf (BGWL)	White leaf	Lee *et al.* (1997); Marcone *et al.* (1997); Arocha *et al.* (2005); Marcone *et al.* (1997); Koh *et al.* (2008)
60.	37.	*Dactyloctenium aegyptium* (Poaceae)	Bunchy shoot disease	Bunchy shoot	Mall *et al.* (2007)
61.	40.	*Dactyloctenium aegyptium* (Poaceae)	Grassy shoot disease	Grassy shoot	Blanche *et al.* (2003)

Table 2. (Continued)

62.	31.	*Dichanthium annulatum* (Poaceae)	White leaf	White leaf	Rao *et al.* (2009)
63.	16.	*Echinochloa colonum* (Poaceae)	Rice yellow dwarf (RYD)	Yellowing	Reddy and Jeyarajan (1990)
64.	9.	*Emelia fosbergii* (Asteraceae)	Lethal yellowing	Yellowing	Brown *et al.* (2008)
65.	32.	*Erigeron bonariensis* (Asteraceae)	Witches'-broom disease	Witches'-broom	Davis *et al.* (1994)
66.	23.	*Erodium cicutarium* (Geraniaceae)	Yellow disease	Yellowing	Graf *et al.* (1978)
67.	38.	*Imperata arundianacea* (Poaceae)	White leaf disease	White leaf	Mall *et al.* (2007)
68.	17.	*Imperata arundinacea* (Poaceae)	White leaf disease	White leaf	Rao *et al.* (1990)
69.	6.	*Medicago sativa* (Fabaceae)	Loofah witches'-broom (LfWB)	Witches'-broom	Lee *et al.* (1993a)
70.	21.	*Ocimum* sp.(Labiatae)	Basil little leaf	Little leaf	Arocha *et al.* (2006)
71.	18.	*Parthenium hysterophorus* (Asteraceae)	Phyllody disease	Phyllody	Padmanabhan (1982)
72.	25.	*Paspalum conjugatum* (Poaceae)	White leaf disease	Bleached leaf	Koh *et al.* (2008)
73.	20.	*Pennisetum purpureum* (Poaceae)	Napier grass stunt disease	Stunting	Nielsen *et al.* (2007)
74.	35.	*Phyllanthus fraternus* (Euphorbiaceae)	Leaf phyllody disease	Leaf phyllody	Mall *et al.* (2007)
75.	8.	*Poa annua* (Poaceae)	Annual blue grass white leaf (ABGWL)	White leaf	Lee *et al.* (1997)
76.	12.	*Polygala mascatense* (Polygalaceae)	Polygala witches'-broom	Witches'-broom	Livingston *et al.* (2006)
77.	22	*Rhynchosia minima* (Fabaceae)	Rhynchosia little leaf	Little leaf	Dabek *et al.* (1983)
78.	30.	*Sasa fortunei* (Poaceae)	Witches'-broom disease	Wwitches'-broom	Zhang *et al.* (2009)
79.	11.	*Senecio jacobaea* (Asteraceae)	Aster Ragwort yellows	Yellowing	Reeder *et al.* (2008)
80.	33.	*Sicana odorifera* (Cucurbitaceae)	Witches'-broom disease	Witches'-broom	Montano *et al.* (2000)
81.	41.	*Sorghum stipoidem*	White leaf disease	White leaf	Blanche *et al.* (2003)
82.	10.	*Synedrella nodiflora* (Asteraceae)	Lethal yellowing	Yellowing	Brown *et al.* (2008)
83.	1.	*Trifolium* sp. (Fabaceae)	Clover Phyllody (Cph)	Phyllody	Lee *et al.* (1992)

84.	3.	*Trifolium* sp. (Fabaceae)	Clover yellow edge (CYE)	Chlorosis, Stunting	Lee *et al.* (1992, 1993)
85.	5.	*Trifolium* sp. (Fabaceae)	Clover proliferation (CP)	Proliferation of axillary shoots	Deng *et al.* (1991); Lee *et al.* (1991)
86.	19.	*Urochloa panicoides* (Poaceae)	Yellowing disease	Yellowing	Muniyappa *et al.* (1982)
87.	27.	*Vernonia cineria* (Asteraceae)	Lethal yellowing	Yellowing	Brown *et al.* (2008)
88.	2.	*Vicia faba* (Fabaceae)	Faba been Phyllody (FBP)	Phyllody	Schneider *et al.* (1995)
89.	43.	*Whiteochloa biciliata* (Poaceae)	White leaf disease	White leaf	Blanche *et al.* (2003)
90.	42.	*Whiteochloa cymbiforms* (Poaceae)	White leaf disease	White leaf	Blanche *et al.* (2003)

Figure 1. Symptoms of phytoplasma on weeds : *(a) witches'-broom on Cannabis sativa, (b) Bermuda grass with white leaves, (c) yellows streaks on leaves of Oplismenus burmanni, (d) pronounced yellowing of Achyranthes aspera, (e) Dichanthium annulatum showing symptoms of extensive chlorosis (f) Crotalaria tetragonaloba with witches'-brooms along with healthy plant.*

4.2 Microscopy

In earlier days, the phytoplasma pathogen were identified on the basis of DAPI staining and electron microscopy detection. Microscopic methods including transmission electron microscopy (TEM) and light microscopy have been used to detect phytoplasmas in weeds in many countries, but most sensitive is the DAPI (DNA-specific-6-diaminido-2-phenylindole) fluorescence microscopy technique (Deeley *et al.*, 1979). However, these techniques require tissue fixation and also these methods are slow and results may be difficult to interpret. Association of phytoplasmas with many weed species were also reported in India on the basis of symptomatology, temporary remission of symptoms after tetracyclin treatments of affected plants and electron microscopy. These reports include , *Cynodon dactylon* (Singh *et al.*, 1978), *Acanthospermum hispidium* (Raju and Muniyappa, 1981), *Parthenium hysterophorus* (Padmanabhan, 1982), *Urochloa panicoides* (Muniyappa *et al.,* 1982), *Imperata arundinacea* (Rao *et al.,* 1990) and *Echinochloa colonum* (Reddy and Jeyarajan, 1990).

Molecular Detection

The identification of phytoplasma disease was previously mostly relied on symptom expression but some times it is too difficult to identify the diseased plant because of absence of specific symptoms on infected plants throughout thelife cycle of the infected plants . For this purpose, an authentic characterization methodology is required for quick and authentic diagnosis of phytoplasmas at early stages of infection. The application of PCR to the diagnosis of phytoplasma-associated weed diseases has greatly facilitated the detection and identification of a wide array of phytoplasmas in hundreds of weed species (Seemüller *et al.*, 1998; Lee *et al.*, 2000; Blanche *et al.*, 2003; Marcone *et al.*, 2004). PCR assays using universal primers are most useful for preliminary diagnosis of phytoplasmal diseases. Several universal and many phytoplasma group-specific primers have been designed for routine detection of phytoplasmas (; Lee and Davis, 1986; Gundersen *et al.*, 1994; Lee *et al.*, 1993b; Seemuller *et al.*, 1994; Kirkpatrick and Smart, 1995; Davis and Sinclair, 1998;; Lee *et al.,* 1998; Harrison *et al.*, 2008; Blanche *et al.,* 2003). Nested-PCR assay increases both sensitivity and specificity and is a valuable technique in the amplification of phytoplasmas from samples in which unusually low titers are present, or substantial inhibitors that may interfere with the PCR efficacy are present (Lee *et al.,* 1993b). In nested-PCR universal primers are used for the preliminary amplification and then followed by second amplification using a second group-specific primers. Therefore, nested-PCR enables the detection of dual or multiple phytoplasmas present in the infected tissues in case of mixed infections (Lee *et al.,* 1993b; Berges *et al.*, 2000; Blomquist and Kirpatrick, 2002; Christensen *et al.*, 2004). For amplification of phytoplasmal ribosomal DNA (rDNA) by PCR assays, the universal phytoplasma primer pairs P1/P7 and P4/P7 (Schneider *et al.,* 1995), P1/P6 (Deng and Hiruki, 1991), and R16F2n/R16R2 (Gundersen and Lee, 1996) are commonly used. Other primers pairs used for amplifying phytoplasma rDNA are listed in Table 3. PCR amplified-products are cloned and sequenced and the sequences obtained are submitted to GenBank. These sequences are then used to determine the genetic relatedness of the phytoplasmas. The taxonomic and molecular identity of phytoplasmas occurring in weed species throughout the world is listed in Table 3.

Table. 3. Taxonomic and molecular classification of phytoplasmas occurring in weed species throughout the world.

SNo.		Plant host	Strain	Phylogenetic group/subgroup	Accession Nos. (16Sr)/ (rp)	Primes used	Reference
91.	24.	Acanthospermum hispidium (Asteraceae)	-	-	-	-	Raju and Muniyappa (1981)
92.	28.	Achyranthes aspera(Amaranthaceae)	-	16SrI	EU573926	P1/P6 and R16F2n/ R16R2	Raj et al. (2008)
93.	4.	Asclepias sp. (Asclepiadaceae)	Milkweed yellows (MWY)	16SrIII-F (X-disease)	-	-	Griffiths et al. (1994) Gunderseon et al. (1994)
94.	14.	Axonopus compressus (Poaceae)	Carpet grass white leaf (CGWL)	-	-	-	Sdoodee et al. (1999)
95.	13.	Brachiaria distachya (Poaceae)	Brachiaria white leaf (BraWL)	16SrXIV-A	-	-	Chen et al. (1972)
96.	15.	Bupleurum falcatum (Apiaceae)	Bupleurum witches'-broom	-	-	-	Shiomi et al. (1983)
97.	26.	Cannabis sativa (Cannabinaceae)	-	16SrI	EU439257	P1/P6 and R16F2n/ R16R2	Raj et al. (2008)`
98.	34.	Cassia italica (Fabaceae)	-	16SrIX	EF666051	P1/P7 and rpL2F3/rp (1)R1A	Khan et al. (2007)
99.	29.	Crotalaria tetragona (Fabaceae)	-	16SrI 16SrIII	FJ185141	R16F2m/ R1-fU5/rU3	Baiswar et al. (2009); Ribeiro et al . (2001)
100.	7.	Cynodon dactylon (Poaceae)	Bermuda grass white leaf (BGWL)	16SrXIV-A	AJ550984 AF248961 AF100412 Y15868 EU032485	P1/P7	Marcone et al. (1997); Lee et al. (1997); Dafalla and Cousin (1988); Schneider et al. (1999); Rao et al. (2007)
101.	31.	Dichanthium annulatam (Poaceae)		16SrXIV	FJ348654	P1/P6 and R16F2n/ R16R2	Rao et al. (2009)
102.	16.	Echinochloa colonum (Poaceae)	Rice yellow dwarf (RYD)	-	-	-	Reddy and Jeyarajan (1990)
103.	9.	Emelia fosbergii (Asteraceae)	-	16SrIV (Lethal yellowing)	EU026214	LY16Sf/L Y16S23r	Brown et al. (2008)

Table 3. (continued)

104.	32.	*Erigeron bonariensis (Asteraceae)*	-	16SrI	-	-	Davis *et al.* (1994)
105.	23.	*Erodium cicutarium* (Geraniaceae)	-	-	-	-	Graf *et al.* (1978)
106.	17.	*Imperata arundinacea* (Poaceae)	-	-	-	-	Rao *et al.* (1990)
107.	6.	*Medicago sativa* (Fabaceae)	Loofah witches'-broom (LfWB), (ArAWB), Alfalfa witches'-broom	16SrVIII-A, 16SrVII-C, 16SrI (Loofah witches'-broom)	L33764/ L27027	P1/P7and R16F2n/ R16R2	Lee *et al.* (1993a)
108.	21.	*Ocimum* sp. (Labiatae)	Basil little leaf	16SrI	-	-	Arocha *et al.* (2006)
109.	18.	*Parthenium hysterophorus* (Asteraceae)	-	16SrI	EU375485	P1/P6 and R16F2n/ R16R2	Padmanabhan (1982); Raj *et al* . (2008)
110.	25.	*Paspalum conjugatum* (Poaceae)	-	-	EU234512	P1/Ptint R16F2n/ R16R2	Koh *et al.* (2008)
111.	20.	*Pennisetum purpureum* (Poaceae)	Napier grass stunt disease	16SrXI	EF012650	-	Nielsen *et al.* (2007)
112.	8.	*Poa annua* (Poaceae)	Annual blue grass white leaf (ABGWL)	16SrXIV-A	-	-	Lee *et al.* (1997)
113.	12.	*Polygala mascatense* (Polygalaceae)	Polygala witches'-broom	16SrII	-	-	Livingston *et al.* (2006)
114.	22.	*Rhynchosia minima* (Fabaceae)	-	-	-	-	Dabek *et al.* (1983)
115.	30.	*Sasa fortunei* (Poaceae)	-	16srI	FJ501958	R16mF2/ R16mR1 and R16F2n/ R16R2	Zhang *et al.* (2009)
116.	11.	*Senecio jacobaea* (Asteraceae)	Aster ragwort yellows	16SrI (Aster yellow)	EU096553	R16F2m/ R1and R16F2n/R 16R2	Reeder *et al.* (2008)
117.	33.	*Sicana odorifera* (Cucurbitaceae)	-	16SrIII	-	-	Montano *et al* . (2000)

118.	10.	*Synedrella nodiflora* (Asteraceae)	-	16SrIV (Lethal yellowing)	EU026213	LY16Sf/L Y16S23r	Brown *et al.* (2008)
119.	1.	*Trifolium* sp. (Fabaceae)	Clover Phyllody (Cph)	16SrI-C (Aster yellow)	L33762	-	Lee *et al.* (1992)
120.	3.	*Trifolium* sp. (Fabaceae)	Clover yellow edge(CYE)	16SrIII-B (X-disease)	L33766/L27091	-	Lee *et al.* (1992; 1993a)
121.	5.	*Trifolium* sp. (Fabaceae)	Clover proliferation (CP)	16SrVI-A (Clover proliferation)	L33761/ L27011	-	Deng *et al.* (1991); Lee *et al.* (1991)
122.	19.	*Urochloa panicoides* (Poaceae)	-	-	-	-	Muniyappa *et al.* (1982)
123.	27.	*Vernonia cineria* (Asteraceae)	CLY	16SrIV	EU057983	P1/P7 and LY16Sf/L Y16-23Sr	Harrison *et al.* (2002); Brown *et al.* (2008)
124.	2.	*Vicia faba* (Fabaceae)	Faba been Phyllody (FBP)	16SrII-C (Peanut witches'-broom)	X83432	-	Schneider *et al.* (1995)

VECTORS OF WEED PHYTOPLASMAS

Till date, very little work has been done on identification of insect vectors especially on phytoplasmas associated with weeds. Weed plants could play a key role in the epidemiology of the disease since they influence the population density of the vectors and act as source of inoculum (Pasquini *et al.*, 2007). "Boir noir" (BN) is an important grapevine disease associated with phytoplasma, it is naturally transmitted by the *Hyalesthes obsoletus* plant hopper vector that complete its life cycle on herbaceous plants specially weeds i.e. *Urtica dioica*; *Calystegia sepium* and *Convolvulus arvensis* (Langer and Maixner, 2004; Maixner *et al.*, 2006). Padmanabhan (1982) reported a phyllody disease of *Parthenium hysterophorus* from South India. The disease is characterized by witches'-broom symptoms and is transmitted by *Hishimonus phycitis*. In Washington another vector *Circulifer tenellus* (Baker) was reported to transmit Columbia basin potato purple top phytoplasma from weeds to other crops (Munyaneza *et al.*, 2007).

GENETIC DIVERSITY

Weed phytoplasmas show a wide geographical distribution (Table 1). So far more than 43 weed species are reported having phytoplasma infections from all over the world. Nucleotide sequence studies have shown that weed-infecting phytoplasmas mainly belongs to five major groups groups 16SrI, 16SrIV, 16SrVIII, 16SrXI and 16SrXIV .(Fig. 3 and 4; Tab. 3). Among them, 16SrI and 16SrXIV phytoplasmas have a more wide occurrence in nature all over the world. In India, phytoplasmas of the 16SrXIV group (= 'Candidatus Phytoplasma cynodontis') are reported as major group occurring on grasses whereas aster yellows (16SrI)group phytoplasmas are reported in in *Cannabis sativa*, *Achyranthes aspera* and *Parthenium hysterophorus* (Rao *et al.*, 2007; 2009; Raj *et al.*, 2008; Mall *et al.*, 2009).

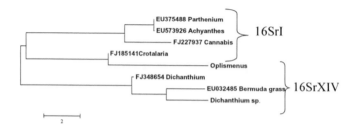

Figure 3. Phylogenetic tree constructed by using MEGA 4.0 version showing relationships among identified weed phytoplasmas in India.

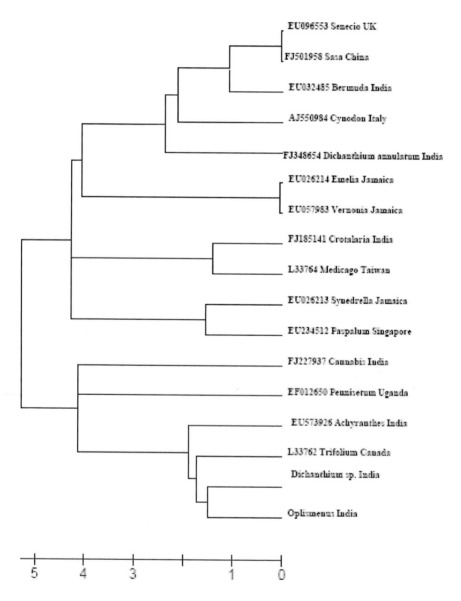

Figure 4. Phylogenetic tree constructed by using neighbour-joining method showing realationships among different phytoplasmas identified in weeds all over the world.

MANAGEMENT

Unfortunately, no effective majors of curing phytoplasma diseases are available at present. The impact of these diseases depends on several factors such as the virulence of strains within a given taxon and their tendency to mutate, the presence and dynamics of the vectors, phytoplasma concentration in both host plants and insect vectors, and environmental conditions and agronomic practices. As a consequence no single control strategy can be adopted. The most important factors to consider before intervening are: disease severity, whether or not to rouge infected plants, rouging strategies, availability of insect vectors, alternative susceptible plant reservoirs and economic impact of the disease (Osler and Carraro, 2004). For the management of these diseases use of disease-free plant propagating material is the most important, because phytoplasmas are primarily spread through grafting of infected plant parts.

Work has also been carried out investigating the effectiveness of <u>plant bodies</u> targeted against <u>phytoplasmas</u>. <u>Tetracyclines</u> are <u>bacteriostatic</u> to phytoplasmas, and inhibit their growth (Singh *et al.,* 1978). Also, Singh and Pathak (1982) reported that tiamulin antibiotic suppressed the symptoms caused by phytoplasma disease associated with *Cynodon dactylon.* However, without continuous use of the antibiotic, disease symptoms will reappear. Thus, tetracycline is not a viable control agent in agriculture, but it is used to protect ornamental coconut trees. It has been long time to know that heat therapy can help in eliminating pathogens such as virus and possibly phytoplasmas from plants and mainly fruit trees (Kunkel, 1936; Kassanism, 1954; Nyland, 1962). For the elimination of phytoplasmas, Laimer and Balla (2003), reported a combination of heat therapy and meristem culture without tetracycline treatment. These strategies may be applied in the case of weed phytoplasmas. Weed phytoplasmas can be best managed by eliminating perennial and biennial weed hosts, rogue out and destroyed symptomatic plants, avoid planting susceptible crops next to crops harbouring phytoplasmas, control of leafhopper vectors in the crop and nearby weeds early in the season (Welliver *et al.,* 1999).

CONLUSIONS

From above study we conclude that the identified phytoplasmas in weeds mainly belongs to 16SrI, 16SrIV, 16SrVIII, 16SrXI and 16SrXIV groups. Among which 16SrI and 16SrXIV phytoplasma groups showed more wide occurrence in nature all over the world. Knowledge of the diversity of phytoplasmas in native and introduced weeds can be better understood by thisreview . The diversity of the potential reservoir of disease has been increased with the discovery of new phytoplasma hosts. The potential reservoir of phytoplasmas in weeds may be even larger than already reported. There is already evidence available, for host plants other than grasses, that phytoplasmas can be detected in plant species before symptoms appear once, that phytoplasmas are not always detectable in every part of a plant, and that the location of phytoplasma in a plant can vary over time.

Even though the weeds identified as phytoplasma hosts often grow abundantly around field crops, the possibilities of transmission of phytoplasmas related to important agricultural, economical and horticultural crops from weed to the agricultural economical crops and vice-versa can not be ignored. This could be because phytoplasmas are able to survive in many potential economical crops (Harrison *et al.,* 2008), or because an insect vector is capable of transmitting phytoplasmas from other weeds to crops which are already known as phytoplasma hosts (; Lee *et al.,* 2004; Saady and Khan , 2006; Harrison *et al.,* 2008). In either case, the chance of transmission in the future seems high, given the large phytoplasma reservoir already revealed, the chance of new phytoplasma strains to evolve (Lee *et al.,* 2000), and the ability of many leafhoppers, the most common insect vectors of phytoplasmas, to migrate long distances (Taylor, 1985) and switch to new host plants (Purcell, 1985).

REFERENCES

[1]   Al-saady, N. A. & Khan, A. J. (2006). Phytoplasmas that can infect diverse plant species worldwide. *Physiol. Mol. Biol. Plant., 12(4):* 263-281.

[2]   Anonymous (1993). International Committee on Systematic Bacteriology, Subcommittee on the Taxonomy of Mollicutes. Minutes of the Interim meetings, 1 and 2 August, 1992, Ames, Iowa. *Int. J. Syst. Bacteriol., 43*: 394-397.

[3]   Anonymous (1997). International Committee on Systematic Bacteriology, Subcommittee on the Taxonomy of Mollicutes. Minutes of the Interim meetings, 12 and 18 August, 1996, Orlando, Florida, USA. *Int. J. Syst. Bacteriol., 47*: 911-914.

[4]   Anonymous (2004). *IRPCM* Phytoplasma/Spiroplasma Working Team – Phytoplasma taxonomy group, '*Candidatus* Phytoplasma', a taxon for the wall-less, non-helical prokaryotes that colonize plant phloem and insects. *Int. J. Syst. Evol. Microbiol, 54*: 1243-1255.

[5]   Arocha, Y., Horta, D., Pinol, B., Palenzuela, I., Picornell, S., Almeida, R. & Jones, P. (2005). First report of Phytoplasma associated with Bermuda grass white leaf disease in Cuba. *Plant Pathol., 54*:233.

[6]   Bai, X., Zhang, J., Holford, I. R. & Hogenhout, S. A. (2004). Comparative genomic identifies genes shared by distantly related insect-transmitted plant pathogenic mollicutes. *FEMS Microbiol. Lett., 235*: 249-258.

[7]   Berges, R., Rott, M. & Seemuller, E. (2000). Range of phytoplasma concentrations in various host plants as determined by competitive polymerase chain reaction. *Phytopathology, 90*: 1145-1152.

[8]   Blanche, K. R., Tran-Nguyen, L. T. T. & Gibb, K. S. (2003). Detection, identification and significance of phytoplasmas in grasses in northen Australia. *Plant Pathol., 52*: 505-512.

[9]   Blomquist, C. L. & Kirkpatrick, B. C. (2002). Identification of phytoplasma taxa and insect vectors of peach yellow leaf roll disease in California. *Plant Dis., 86*: 759-63.

[10] Bove, J. M. (1984). Wall-less prokaryotes of plants. *Ann. Rev. Phytopathol., 22*: 361-396.

[11] Brown, S. E., Been, B. O. & McLaughlin, W. A. (2008). First report of the presence of the leathal yellowing group (16Sr IV) of phytoplasma in the weeds Emelia fosbergii and Synedella nodiflora in Jamaica. *Plant Pathology*: http: // www.bspp.org.uk / ndr / jan 2008/2007 - 75. asp.

[12] Chen, T. C., Lee, C. S. & Chen, M. J. (1972). Mycoplasmalike organisms in *Cynodon dactylon* and *Brachiaria distachya* affected by white leaf disease. *Rep. Taiwan Sugar Exp. Stn., 56*: 49-55.

[13] Choi, Y. M., Lee, S. H., Lee, E. K. & Kim, J. S. (1985). An investigation of underscribed witches'broom symptom caused by MLO on *Bupleurum falcatum*, *Cnidium officinale* and *Plantago asiatica* in Korea. *Korean J. Mycol., 13 (1)*: 49- 51.

[14] Christensen, N. M., Axelsen, K. B., Nicolaisen, M. & Schulz, A. (2004). Phytoplasmas and their interactions with hosts. *Trends Plant Sci., 11*: 526-535.

[15] Cronje, P., Dabek, A. J., Jones, P. & Tymon, A. M. (2000a). First report of a phytoplasma associated with a disease of date palms in North Africa. *Plant Pathol., 49*: 801.

[16] Cronje, P., Dabek, A. J., Jones, P. & Tymon, A. M. (2000b). Slow decline a new disease of mature date palms in North Africa associated with a phytoplasma. *Plant Pathol., 49*:804.

[17] Dabek, A. J. (1983). Leaf hopper transmission of Rhynchosia little leaf, a disease associated with MLO in Jamaica. *Phytopathol..Z., 107 (4)*: 345-361.

Dafalla, G. A. & Cousin, M.T. (1988). Fluorescence and electron microscopy of *Cynodon dactylon* affected with a white leaf disease in Sudan. *J. Phytopathol., 122:* 25-34.

[18] Davis, R. E., Dally, E. L., Lee, I. M., Resende, R. O. & Kitajima, E. W. (1994). Detection and characterization of mycoplasmalike organisms (MLOs) in diverse plant species in Brazil. *Fitopatol. Brasil., 19*:340.

[19] Davis, R. E. & Dally, E. L. (2001). Revised subgroup classification of group 16SrV phytoplasmas and placement of flavescence dorée-associated phytoplasmas in two distinct subgroups. *Plant Dis., 85*: 790-797.

[20] Davis, R. E. & Sinclair, W. A. (1998). Phytoplasma identity and disease etiology. *Phytopathology, 88*: 1372-1376.

[21] Deeley, J., Stevens, W. A. & Fox, R. T. V. (1979).Use of Dienes stain to detect plant diseases induced by MLO. *Phytopathology 69*:1169-1171.

[22] Deng, S. & Hiruki, C. (1991). Amplification of 16S rRNA genes from culturable and non- culturable mollicutes. *.J. Microbiol. Methods, 14*: 53-61.

[23] Doi. Y., Teranaka, M., Yora, K. & Asuyama, H. (1967). Mycoplasma or PLT group-like micro-organisms found in the phloem elements of plants infected with mulberry dwarf, potato witches, broom, aster yellow or Paulownia witches' broom. *Ann. Phytopath. Soc. Japan, 33*:259-266.

[24] Gibb, K. S., Schneider, B. & Padovan, A. C. (1998). Differential detection and relatedness of phytoplasmas in papaya. *Plant Pathol., 47*: 325-332.

[25] Gibb, K., De. La Rue, S., Eichner, R. & Sdooder, R. (1997) . Detection and identification of phytoplasmas associated with white leaf and stunting disease in Australian Grasses. *Australian Plant Pathology Society.* 11th Bienial Conf. (Abs) Perth, 29 Sep.–20 Oct., 1997.

[26] Graf, M. E., Ehrenfeld, R. & Davis, R. E. (1978). Stereo electron microscopy of MLO in *Erodium cicutarum* with yellows disease symptoms. *Plant Dis. Rep., 62(6):* 535-538.

[27] Gundersen, D. E. & Lee, I.-M. (1996). Ultrasensitive detection of phytoplasmas by nested-PCR assays using two universal primer pairs. *Phytopath. Medit. 35* : 144-151.

[28] Gundersen, D. E., Lee, I-M., Rehner, S. A., Davis, R. E. & Kingsbury, D. T. (1994). Phylogeny of mycoplasmalike organisms (phytoplasmas): a basis for their classification. *J. Bacteriol., 176*:5244-5254.

[29] Harrison, N. A. & Oropeza, C. (2008). Coconut Lethal Yellowing, pp. 219-248. In: Characterization, Diagnosis and Management of Phytoplasmas (eds. N. A. Harrison, G. P. Rao & C. Marcone). Studium Press LLC, U.S.A.

[30] Ho, K., Tsai, C. & Chung, T. (2001). Organization of ribosomal RNA genes from a loofah witches' broom phytoplasma. *UNA and Cell Biol., 20* : 115-22.

[31] Jung, H.-Y., Sawayanagi, T., Wongkaew, P., Kakizawa, S., Nishigawa, H., Wei, W., Oshima, K., Miyata, S.-i., Ugaki, M., Hibi., T. & Namba, S. (2003). 'Candidatus Phytoplasma oryzae', a novel phytoplasma taxon associated with rice yellow dwarf disease. *Int. J. Syst. Evol. Microbiol., 53*: 1925-1929.

[32] Kassanis, B. (1954). Heat therapy of virus infected plants. *Ann. Appl. Biol., 41(3)*: 470-474.

[33] Kenyon, L., Harrisson, N. A., Ashburner, G. R., Boa, E. & Richardson, P. A. (1998). Detection of a pigeon pea witches'-broom - related phytoplasma in trees of Gliricidia sepium affected by little – leaf disease in Central America. *Plant. Pathol., 47*: 671- 680.

[34] Khan, A.J., Al-Subhi, A.M., Calari, A., Al-Saady, N.A. & Bertaccini, A. (2007). A new phytoplasma associated with witches'-broom of *Cassia italica* in Oman. *Bull. Insectology 60(2)*:269-270 .

[35] Khan, A. J., Botti, S., Al-Subhi, A. M., Gundersen-Rindal, D. E. & Bertaccini, A. (2002a). Molecular identification of a new phytoplasma strain associated with alfalfa witches' broom in Oman. *Phytopathology, 92*: 1038-1047.

[36] Kirkpatrick, B. C., Smart, C., Gardner, S. Gao, J-L., Ahrens, U., Maurer, R., Schneider, B., Lorenz, K-H., Seemuller, E., Harrison, N., Namba, S. & X. Daire. (1994). Phylogenetic relationships of plant pathogenic MLOs established by 16/ 23S rDNA spacer sequences. *IOM Letters, 3* : 228-229.

[37] Kirkpatrick, B. C. & Smart, C. D. (1995). Phytoplasmas: can phylogeny provide the means to understand pathogenicity, pp. 187-212. In: Advances in Botanical Research Vol 21 (eds.: J. H. Andrews & I. C. Tommerup). Academic Press, NY, USA.

[38] Koh, L. H., Yap, M. L., Yik, C. P., Niu, S. N. & Wong, S. M. (2008). First report of Phytoplasma Infection of Grasses in Singapore. *Plant Dis., 92*: 317.

[39] Kunkel, L. O. (1936). Heat treatments for the cure of yellows and other virus diseases of peach. *Phytopathology, 26*: 809-830.

[40] Kuske, C. R. & Kirkpatrick, B. C. (1992). Phylogenetic relationships between the western aster yellows amycoplasmalike organism and other prokaryotes established by 16S rRNA gene sequence. *Int. J. Syst. Bacteriol., 42*: 226-233

[41] Laimer, M. & Balla, I. (2003). Méthodes rapides et fiables pour la détection et l'élimination des Phytoplasmes chez les arbres fruitiers. *Fruit Belge, 505*: 157 – 161.

[42] Lee, I. M. & Davis, R. E. (1992). Mycoplasmas which infect plants and insects, pp 379-390. In: Mycoplasmas: Molecular Biology and Pathogenesis (eds.: J. Maniloff, R. N. McElhaney, L. R. Finch & J. B. Baseman). American Society for Microbiology, Washington, DC, USA.

[43] Lee, I. M., Davis, R. E. & Gundersen-Rindal, D. E. (2000). Phytoplasma: phytopathogenic mollicutes. *Annu. Rev. Microbiol., 54*: 221-255.

[44] Lee, I. M., Gundersen-Rindal, D. E., Davis, R. E. & Bartoszyk, M. (1998). Revised classification scheme of phytoplasmas based on RFLP analyses of 16S rRNA and ribosomal protein gene sequences. *Int. J. Syst. Bacteriol, 48*: 1153-1169

[45] Lee, I. M., Hammond, R. W., Davis, R. E. & Gundersen, D. E. (1993b). Universal amplification and analysis of pathogen 16S rDNA for classification and identification of mycoplasmas like organisms. *Phytopathology, 83* : 834-842

[46] Lee, I. M., Zhao, Y., Davis, R. E., Wei, W. & Martini, M. (2007). Prospects of DNA – based systems for differentiation and classification of phytoplasmas. *Bull. Insectology, 60 (2)*: 239 – 244.

[47] Lee, I.-M., Pastore, M., Vibio, M., Danielli, A., Attathorn, S., Davis, R. E. & Bertaccini, A. (1997). Detection and characterization of a phytoplasma associated with annual blue grass (*Poa annua*) white leaf disease in Southern Italy. *Eur. J.Plant Pathol., 103*:251-254.

[48] Lee, I-M. & Davis, R.E. (1986) Prospects for *in vitro* culture of plant-pathogenic mycoplasmalike organisms. *Ann. Rev. Phytopathol., 24*: 339-354.

[49] Lim, P. O. & Sears, B. B. (1989). 16S rRNA sequence indicates that plant-pathogenic mycoplasmas like organisms are evolutionarily distinct from animal mycoplasmas. *J. Bacteriol., 171*: 5901-5906.

[50] Livingston, S., Al-Azri, M.O., Al-Saady, N. A., Al-Subhi, A. M. & Khan, A. J. (2006). First report of 16Sr DNA II group phytoplasma on *Polygala mascatense*, a weed in Oman. *Plant Dis., 90(2):* 248.

[51] Mall, S. (2009). Characterization of Phytoplasma Associated With Weeds in Eastern Uttar Pradesh .D.D.U. Gorakhpur University, Gorakhpur-273009. U.P., India.

[52] Mall, S., Rao, G. P. & Upadhyaya, P. P. (2009). Molecular characterization of Aster yellows (16SrI) group phytoplasma infecting *Cannabis sativa* in Eastern Uttar Pradesh. *Indian Phytopathol.*, (in press).

[53] Mall, S., Chaturvedi, Y., Singh, M., Upadhyaya, P. P., & Rao, G. P.(2007). Molecular characterization of Phytoplasmas infecting weeds in Eastern Uttar Pradesh, India. *Indian J. Plant Pathol., 25 :(1&2):* 77-80

[54] Marcone, C. & Rao, G. P. (2008). '*Candidatus* Phytoplasma cynodontis': The causal agent of Bermuda Grass White Leaf Disease, pp. 353-364. In: Characterization, Diagnosis and Managemen of Phytoplasmas (eds.: N. A. Harrison, G. P. Rao & C. Marcone). Studium Press LLC, U.S.A..

[55] Marcone, C., Neimark, H. Ragozzino, A., Lauer, U. & Seemuller, E. (1999). Chromosome sizes of phytoplasmas composing major phylogenetic groups and subgroups. *Phytopathology, 89*: 805-810.

[56] Marcone, C., Ragozzino, A. & Seemuller, E. (1997). Detection of Bermuda grass white leaf disease in Italy and genetic characterization of the associated phytoplasma by RFLP analysis. *Plant Dis., 81:*862-866.

[57] Marcone, C., Schneider, B. & Seemuller, E. (2004). '*Candidatus* Phytoplasma cynodontis', the phytoplasma associated with Bermuda grass white leaf disease. *Int. J. Syst. Evol. Microbiol, 54*:1077-1082

[58] Marwitz, R. (1990). Diversity of yellows disease agents in plant infections. *Zbl. Bakt. Suppl., 20* : 431-434.

[59] McCoy, R. E., Caudwell, A., Chang, C. J., Chen, T. A., Chiykowski, L. N., Cousin, M. T., Dale, J. L., deLeeuw, G. T. N., Golino, D. A., Hackettt, K. J., Kirkpatrick, B. C.,

Marwitz, R., Petzold, H., Sinha, R. C., Suguira, M., Whitcomb, R. F., Yang, I. L., Zhu, B. M. & Seemuller, E. (1989). Plant diseases associated with mycoplasma-like organisms, pp. 545-640. In: The Mycoplasmas Vol. V (eds.: R. F. Whitcomb & J. G. Tully).. Academic Press, San Diego, USA.

[60]   Montano, H.G., Davis, R. E., Dally, E. L. Pimentel, J. P. & Brioso, P.S.T.(2000). Identification and phylogenetic analysis of a new phytoplasma from diseased chayote in Brazil . *Plant Dis., 84*: 429-436.

[61]   Muniyappa, V., Raju, B. C. & Nyland, G. (1979). White leaf disease of Bermuda grass in India. *Plant Dis. Rep., 63(12)*: 1072-1074.

[62]   Muniyappa,V., Rao, M. S. & Govindu, H. C. (1982). Yellowing disease of *Urochloa panicoides* Beauv. *Curr. Sci. 51(8):* 427-428.

[63]   Munyaneza, J. E., Crosslin, J. M. & Upton, J. E. (2007). Beet leafhopper (Hemiptera: Cicadellidae) transmits the Columbia basin potato purple top phytoplasma to potatoes, beets and weeds. *J. Econ. Entomol. , 94(2);* 268-272.

[64]   Murray, R. G. E. & Schleifer, K. H. (1994). *Taxonomic notes a proposal for recording the properties of putative taxa of prokaryotes. Int. J. Syst. Bacteriol., 44*: 174–176.

[65]   Namba, S., Katao, S., I wanami, S., Oyaizu, H., Shiozawa, H. & Tsuchizaki, T. (1993). Detection and differentiation of plant pathogenic MLOs using PCR. *Phytopathology, 83*:786-781

[66]   Nyland, G. (1962). Thermotherapy of virus infected fruit trees. Proceedings of the Fifth European Symposiuum on Fruit Tree Virus Diseases, Bologna pp. 156-160.

[67]   Oshima K., Kakizawa S., Nishigawa H., Jung H. Y., Wei W., Suzuki S., Arashida R., Nakata D., Miyata S., Ugaki M. & Namba S. (2004*).* Reductive evolution suggested from the complete genome sequence of a plant-pathogenic phytoplasma. *Nature Genet. 36:* 27-29.

[68]   Osler, R. & Carraro, L. (2004). Gli scopazzi del melo. *Inf. Fitopatol., 5*: 3-6.

[69]   Padmanabhan, C. (1982). A phyllody disease of *Parthenium hysterophorus* L. *Current Rearch,* Agri-Coll. Res. Inst., Coimbatore 641003, India. Plant Pathology Circular No. 82 Pennsylvania Dept. of Agriculture Spring.

[70]   Purcell, A. H. (1985). The ecology of bacterial and mycoplasma plant diseases spread by leafhoppers and planthoppers, pp. 351–380. In: The Leafhoppers and Planthoppers (eds.: L. R. Nault & J. G. Rodriguez). John Wiley & Sons,New York, NY , USA..

[71]   Raj, S. K., Snehi, S. K., Khan, M. S. & Kumar, S. (2008). '*Candidatus* Phytoplasma asteris' (group 16SrI) associated with a witches'-broom disease of *Cannabis sativa* in India. *Plant Pathology*: http: // www.bspp.org.uk / ndr / jan 2008/2007 - 75. asp.

[72]   Rao ,G. P., Mall, S., Singh , M. & Marcone,C. (2009). First report of a '*Candidatus* Phytoplasma cynodontis'-related strain (group 16SrXIV) associated with white leaf disease of *Dichanthium annulatum* in India. *Australas. Plant Dis. Notes, 4,* 1–3.

[73]   Rao, G. P. & Singh, H N. (1990). Occurrence of grassy shoot disease pathogen (MLO) of sugarcane on *Imperata arundinacea* Cyrill in India. *Nat. Acad. Sci. Letters, 13(11):* 403.

[74]   Rao, G. P., Raj, S. K., Snehi, S. K., Mall, S., Singh, M. & Marcone, C. (2007). Molecular evidence for the presence of '*Candidatus* Phytoplasma cynodontis', the Bermuda grass white leaf agent in India. *Bull. Insectology 60(2):* 145-146.

[75]   Reddy, A. V. & Jeyarajan, R. (1990). Rice yellow dwarf- host range studies and its effect on growth of weed host. *Indian J. Mycol. and Plant Pathology, 20 (1)*: 26 – 29.

[76]   Reeder, R. & Arocha, Y. (2008). 'Candidatus Phytoplasma asteris' identified in Senecio jacobaea in the United Kingdom. Plant Pathology : http://www.bspp.org.uk/ndr / jan 2008/2007 - 91. asp.

[77]   Sarindu, N. & Clark, M. F. (1993). Antibody production and identity of MLOs associated with Sugar cane whiteleaf disease and bermuda – grass white leaf disease from Thailand. Plant Pathol., 42: 396-402.

[78]   Schneider, B., Padovan, A., De La Rue, S., Eichner, R., Davis, R., Bernuetz, A. & Gibb, K. S.(1999). Detection and differentiation of phytoplasmas in Australia: an update. Aust. J. Agric. Res., 50: 333-342.

[79]   Schneider, B., Seemüller, E., Smart, C. D. & Kirkpatrick, B. C. (1995). Phylogenetic classification of plant pathogenic mycoplasma-like organisms or phytoplasmas, pp. 369-380. In: Molecular and Diagnostic Procedures in Mycoplasmology, Vol. I. Molecular Characterization (eds.: S. Razin, & J. G. Tully). Academic Press Inc., San Diego, California, USA.

[80]   Sdoodee, R., Schneider, B., Padovan, A.C. & Gibb, K.S. (1999). Detection and genetic relatedness of phytoplasmas associated with plant diseases in Thailand. J. Biochem. Mol. Biol. and Biophys., 3: 133-140.

[81]   Seemuller, E., Marcone, C., Lauer, U., Ragozzino, A. & Goschl, M. (1998). Current status of molecular classification of the Phytoplasmas. J. Plant Pathol., 80:3-26

[82]   Seemuller, E., Schneider, B., Maurer, R., Ahrens, U., Daire, X., Kison, H., Lorenz, K. H., Firrao, G., Avinent, L.,Sears, B. B. & Stackebrandt, E. (1994). Phylogenetic classification of phytopathogenic mollicutes by sequence analysis of 16SrDNA. Int. J. Syst. Bacteriol, 44: 440-446.

[83]   Shmuel Razin. (2007). Molecular biology and genomics of Mollicutes. Bull. Insectology, 60 (2): 101–103.

[84]   Sinclair, W. A., Griffiths, H. M. & Davis, R. E. (1996). Ash yellows and lilac witches'-broom: phytoplasmal diseases of concern in forestry and horticulture. Plant Dis., 80: 468-475.

[85]   Singh, U. P., Sakai, A. & Singh, A. K. (1978). White leaf disease of Cynodon dactylon Pers., a mycoplasmal disease in India. Experientia, 34 (11): 1447-1448

[86]   Smrze, J., Ulrychova, M. & Jokes, M. (1981). Alfalfa witches' broom in Czechoslovakia and demonstration of MLO in the affected plants. Biol. Plant. 23(5): 381-383.

[87]   Taylor, R. A. J. (1985). Migratory behaviour in Auchenorrhyncha, pp. 259-288. In: The Leafhoppers and Planthoppers (eds.: L. R., Nault & J. G. Rodriguez). . John Wiley & Sons. New York, NY, USA.

[88]   Tran-Nguyen, L., Blanche, K. R., Egan, B. & Gibb, K. S. (2000). Diversity of phytoplasmas in northern Australian sugar cane and other grasses. Plant Pathol., 49:666-679.

[89]   Tsai, J. H. (1979). Vector transmission of mycoplasmal agents of plant diseases, pp. 265-307. In: The Mycoplasmas, Vol. III. (ed.: R. E. McCoy). Academic Press, San Diego, USA..

[90]   Viswanathan, R. (1997). Detection of phytoplasmas associated with grassy shoot disease of sugarcane by ELISA techniques. Z. PflKranKh. PflaShutz, 104: 9-16.

[91]   Welliver, R. (1999). Diseases Caused by Phytoplasmas. [Vol. 25, No.1] 1999 Bureau of Plant Industry.

[92] Zahoor, A., Bashir, M., Nakashima, K., Mitsueda, T. & Murata, N. (1995). Bermuda grass white leaf caused by phytoplasma in Pakistan. *Pak. J. Bot.*, *27*:251 – 252.

In: Recent Trends in Biotechnology and Microbiology
Editors: Rajarshi Kumar Gaur et al., pp. 109-115

ISBN: 978-1-60876-666-6
2010 Nova Science Publishers, Inc.

Chapter 9

AN OVERVIEW OF GENE SILENCING AND ITS APPLICATIONS

Richa Raizada, Pradeep Sharma, Rajneesh Prajapati, Abhinav Tyagi, M.S. Rathore, K.P.Sharma and R.K.Gaur*

Department of Science, Faculty of Arts, Science and Commerce, Mody Institute of Technology and Science, Lakshmangarh, Sikar, Rajasthan, India

ABSTRACT

The ability of suppress transiently mRNA accumulation of specific genes in a high-throughput is a powerful tool in a genomics-scale approach to assign biological function to uncharacterized genes. Virus induced gene silencing, or VIGS, is a method o transiently interrupt gene function through RNA interference. The exact mechanism by which VIGS operates is still unclear. It is known, however, that this approach harnesses the plants natural ability to suppress the accumulation of foreign RNAs by an RNA-mediated defense mechanism against plant virus. The systemic signal by which this mechanism is induced is unknown, but it is thought to involve an RNA component. Recently it was shown that inoculation of transcript from cloned viruses capable of expressing host sequences in plants led to silencing of the homologous host gene. The infected plants displayed a phenotype representative of the loss of function of the host gene, and not of virus infection.

INTRODUCTION

RNAi has evolved into a powerful technique to silence gene expression in cells. It allows researchers to study molecular effects of modulating expression at the level of individual genes. This degree of precision can now be accomplished without the time consuming efforts previously dedicated to the construction of single gene knock-outs or dominant negative expressing cell lines.

RNA interference (RNAi), i.e. gene silencing, or gene expression down-regulation is the process whereby a double-stranded RNA (dsRNA) is induces the homology despondent degradation of cognate mRNA. When dsRNA is introduced into cell, an RNA silencing complex (RISC) is assembled. RISC serves as cellular machinery that is responsibly for the specific mRNA degradation. These results are subsequent reduction of the specific proton translated from appropriate mRNA. Short RNA duplexes (~21 nucleotide) called small interfering RNA (siRNA), have become the major tool for induction of gene silencing. With the human genome mapped and sequenced, attempts are currently made to manipulate the expression of genes involved in viral diseases, carcinogenesis and in other disorders with the air of developing novel therapies.

MECHANISM

There are at least three RNA silencing pathways of silencing specific gene in plants. In these silencing signals can be amplified and transmitted between cells and may even be self-regulated by feedback mechanisms. Diverse biological roles of these pathways have been established, included defense against viruses, regulation of gene expression and the condensation of chromatin into heterochromatin.

The first of the three is cytoplasmic siRNA silencing. The pathway may be important in virus-infected plant cells here the dsRNA could be a replication intermediate or a secondary-structure feature of single-stranded viral RNA. In case of plant DNA viruses, the dsRNA may be formed by annealing of overlapping complementary transcripts. The originally described recovery from tobacco ringspot virus and many examples of transgenic silencing in plants and animals are probably manifestations of cytoplasmic RNA silencing.

The second pathway is the silencing of endogenous messenger RNAs by mi RNAs. These miRNA negatively regulate gene expression by base pairing to specific mRNAs, resulting in either RNA cleavage of arrest of protein translation. Like is RNA, the miRNA are short 21-24 nucleotide RNAs derived by DICR cleavage of a precursor the protype miRNA in plants were identified as a subset of the short RNA population with the molecular characteristic of the heterochronic RNAs (Hutvagner et al.,2001; Batel, 2004; Lee et al., 2002; Ketting et al., 2001).

The third pathway of RNA silencing in plants is associated with methylation and suppression of transcription. The first evidence for this type of silencing was the discovery in plant that trasgene and viral RNA guide DNA methlation to specific nucleotide sequences.

The RNAi tools could be based on differential exploitation of the insect metabolism, manipulation of the insect immune and endocrine systems to weaken insects defenses and shorten its development; contributing to diminish insecticide abuse and environmental contamination by reducing the extent of non-specific insecticide applications and by improving Integrated Pest Management technique. Among many other possibilities, we could identify several leads;

HOST-PATHOGEN INTERACTION

Microorganisms that invade invertebrate hosts are initially recognized by the innate immune system through pattern-recognition receptors (PRRs). In D. *melanogaster* this response, triggered by pathogen-associated molecular pattern recognition, involves activation of the Toll, Imd and Hop signaling pathway that rapidly induce the expression of a variety of overlapping and unique genes involved in the immune responses. RNAi HTS has been utilized to characterize the Drosophila innate immune responses and can be further implemented in identifying crucial genes in the immune signaling cascades of insects, that their inhibition could weaken the insect response to specific pathogens (e.g. bacteria, fungi and viruses), thus enabling development pf novel strategies for pathogen-mediated pest control. Possible readouts could include specific antigens of the pathogen detected by monoclonal antibodies, pathogen-mediated reported gene-expression, pathogen mediated cell-death, etc. RNAi HTS could be implemented top define insect intracellular responses to virus infection, such as to picornalike viruses infecting beneficial insects like bees or cyupoviruses that could be implemented ion pest control, and specific host responses to viral infection on different scenarios and validation of the involved genes could be performed in their corresponding host.

NOVEL INSECTICIDES

Novel targets will be useful for rational insecticide design strategies. Most neuropeptide and protein hormone receptors belong to the large superfamily of G-protein-coupled receptors (GPCRs) that turn many imported processes such as development, reproduction, homeostasis and behavior when activated by their corresponding ligands. D.*melanogaster* RNAi HTS can be designed to identify signaling pathways triggered by binding of selected neuropeptides and hormones to specific receptors that could be further exploited as novel targets for insect control. For example, the molting cycle is a hallmark of insects; however, neither the endocrine control of molting via size, stage, and nutritional inputs nor the enzymatic mechanism for synthesis and release of the exoskeleton is well understood. A genome-wide RNA-interference screen could contribute to identify endocrine and enzymatic regulators of molting in Drosophila and further identification of orthologous in insects of agricultural economic relevance. This could include transcription factors, secreted peptides, transmembrane proteins, and extracellular matrix enzymes essential for molting. Inactivation of particular genes could reveal regulatory networks that might couple the expression of genes essential for molting to endocrine cues.

INSECTICIDE RESISTANCE

D. *melanogaster* resistance to classical insecticides seems to occur by the same mechanisms as in related insect species. Insecticide resistance can emerge when expression of detoxifying genes increases, or through mutations in genes that encode the protein targeted by insecticides. RNAi HTS to tudy the genetic basis of resistance to neonicotinoids and insect growth regulators in files utilizing as readouts

Fluorescence-based   assays   for   genes   encoding   detoxification   enzymes (Carboxylesterases, Cytochrome P450s and Glutathione-s-transferases) could contribute to reveal the molecular basis of the insecticide sensitivity and resistance.

The Cry family of Bacillus thuuringiensis insecticidal proteins is widely utilized in classic and transgenic approaches to control insect pests. Cry toxin resistance has been associated with deficient protoxin activation by host proteases, and defective Cry toxin-binding cell surface molecules, such as cadherins, aminopeptidases and glycolipids. Understanding Cry toxin resistance at the molecular level is critical to the long-term utilization of Cry proteins may also shed light on basic mechanisms used by other bacterial toxins that target specific organisms or cell types. The use of *D.melanogaster* RNAi HTS in conjunction with insect genomics could be a useful tool for broadening our understanding of Cry toxin resistance. Esterases, aminopeptidases and Cry-mediated cell death could be implemented as possible readouts in the screen.

BIOLOGICAL CONTROL OF TEPHRITIDAE (TRUE FRUIT FLIES)

True fruit flies (like *Ceratitis capitata*) are major agricultural pests in many areas of the world. The main non-polluting mean of control relies on the release of sterile males obtained by gama irradiation. Howeve, those males show reduced vitality after their release as compared to wild type males. Thus, a more targeted approach to chivew male sterility could highly improve the utilization of sterile *Tephritide* file in agricultural application. Functional RNAi screens can be designed to discover genes that participate in the germiline sex-determination pathway or that are aimed at perturbing regulatory networks resulting in male aterility. For instance, it is well established that HID as well as the caspases DRONE, DRICE and DCPI play a role in spermatid differentiation. The pathway and genes that drive caspase activation in these non-apoptotic contests are not known. RNAi acreens relying on caspase activity as a read-out will be highly informative in identifying such genes.

APPLICATION OF RNAI FOR CROP IMPROVEMENT

Directed by T. J. Higgins (Current publications by CSIRO), scientists are CSIRO, in Australia, have played a pioneering role in demonstrating that RNAi technology may be used for such applications as gene silencing thereby generating improved crop varieties in terms of disease-, insects resistance, enhancing nutritional qualities, and much more. To facilitate gene silencing through RNAi they also developed several versions of the pHannibal plasmid vectors from the original constructs. The above scientists have shown that by replacing the loop in hpRNA with an intron, the efficiency of gene silencing can be enhanced from about 50% to nearly 100%. The vector sequences for pHannibal and pKannibal are available in Genbank as well as a published paper by Wesley *et al.* (2001). Scientists all over the world working with RNAi will benefit form their finding and the vectors (pHannibalm pKannibal) which are available free of charge for academic research. *BaycerCrop* Science has acquired an exclusive worldwide license to develop, market, al sell selected crop plant varieties in which the RNAi technology has been successfully applied by the CSIRO scientists. Using this technique this group has developed varieties of barley that are resistance to BYDV (barely yellow dwarf virus) (Wang *et al.* 2000). Their results showed that the barely plants developed through RNAi technology are resistance to viral infection while the control plants became infected with the yellow dwarf virus.

Kusabaand his team (Kusuba *et al.* 2003) have recently made significant contribution by applying RNAi to improve rice plants. They were able to reduce the level of glutenin and produced a rice variety called *LGC-1* (low glutenin content 1). The low glutenin content was a relief to the kidney patients unable to digest glutenin. The trait was transmitted for a number of generations. They showed that the procedure may apply to both monogenic and polygenic agronomic characters. They advocated the use of either a weak promoter to regulate the level of expression of dsRNA or the use of sequence with various homologies to the target gene. Since the use of a weak promoter reduces the frequency of suppression rather than weakens it, they favor the latter method. To reduce the level of suppression, they recommend the use of both closely- or distantly related species that bear various degrees of homologies to the target gene. Close homology between the target gene and the host plant would enable each resulting siRNA to cleave the target mRNA.

PROSPECTS OF UTILIZING TECHNIQUES

In Ethiopia, Bangladesh and India, the people in the lower socioeconomic class use a leafy vegetable known as *Lathrus sativus.* It is a leguminous crop and contains a neurotoxin called B-oxalyaminoalanine-L-alanine (BOAA) (Spencer *et al.* 1986). People consuming this vegetable suffer from a paralytic disease called, lathyrism. The disease paralyses people bothe temporarily and permanently, however the effects can be somewhat reduced if the plants is boiled prior to consumption. Paralysis in the limbs is known symptoms of BOAA, yet people still consume this vegetable in times of famine. This species is remarkably suited to grow in marginal and inhospitable land without irrigation, fertilizer, and pesticides. It flourishes also times of devastating flood and drought. When no other food crop survives. This is an instance where RNAi technology can be used to silence the gene9s) sresponsible for production of BOAA. There may be one difficulty; in that the BOAA genes may be linked to genes, which confer to this unique crop or impart drought and flood tolerance. Bringing down the level of BOAA to a safe concentration, rather than totally silencing the concerned genes, may overcome this obstacle.

Another instance where RNAi may be fruitfully applied is in the production of banana varieties resistance to the Banana Bract Mosaic Virus (BBrMV), currently devastating the banana population in Southeast Asia and India (Rodoni *et al.* 1999). In certain years, the BBrMV infects banana plants destroying the fruit producing bract region, rendering them useless to farmers. The virus is spread by small plant eating insects called aphids, as well as through infected plant materials. The problem is further compounded when further banana crops are raised ion the infected field because the infection spreads from the previous diseased crop. However, by carefully designing an RNAi vector aimed at silencing the Coat Protein (CP) region of the virus, scientist may be able to develop of the different strains of virus is highly conserved and as such silencing of region of the different strains of virus is highly conserved and as such silencing of this gene in other varieties of banana will not pose a problem. Another novel approach here would be to utilize an inducible promoter system in order that dsRNA is produced only upon infectin and not constitutively.

A possible application of RNAi involves the down regulation of key enzyme in the biosynthetic pathway of lignin in the two economically important *Corchours* species, namely, *C. capsularis* and *C. olitorius.* The enzyme 4-coumarate: CoA ligase (4-Cl) is on of the key enzymes in the early stages of lignin biosynthesis. This makes it a promising target for regulating the quantity of lignin, produced in the jute plat. The present quantity of lignin in the commercial varieties of jute increases the cost of pulp production for manufacture of high quality paper. Hence, reduction in the lignin content will be welcome to the paper industry. With the availability of the sequence of the 4-Cl gene, it would be possible to create a transgenic jute variety expressing the RNAi construct to down regulate the quantity of 4-Cl mRNA thereby reducing the lignin production. With this approach, it would also be possible to vary the quantity of lignin synthesis by the help of different promoters and altering the length of interfering RNA. Thus RNAi technology may prove to be powerful molecular tool by generating jute varieties with low lignin content, allowing for easier, environmentally friendly and cost effective processing of fiber for the production of varieties economically important commodities such as high quality paper and cloth.

REFERENCES

[1]   Batel, D.P (2004). *MicroRNAs: genomics, biogenesis, mechanism, and function.* Cell. 116: 281.

[2]   Hutvagner, G., McLachlan, J., Pasquinelli, A.E., Balint, E., Tuschl, T. & Zamore, P.D. (2001). A cellular function for the RNA interference enzyme Dicer in the maturation of the let-7 small temporal RNA. *Science.* 293: 834.

[3]   Ketting, R.F., Fischer, S.E., Bernstein, E., Sijen, T., Hannon, G.J. & Plasterk, R.H. (2001). Dicer functions in RNA interference and in synthesis of small RNA involved in developmental timing in *C. elegans. Genes Dev.* 15: 2654.

[4]   Kusaba, M., Miyahara, K., Lida, S., Fukuoka, H., Takario, T., Sassa, H., Nishimura, M. & Nishio, T. (2003). Low glutenin content 1: a dominant mutation that suppresses the glutenin multigene family via RNA silencing in rice. *Plant Cell,* 15: 1455-1467

[5]   Lee, Y., Jeon, K., Lee, J.T., Kim, S. & Kim, V.N. (2002). MicroRNA maturation: stepwise processing and subcellular localization. *EMBO J.* 21: 4663.

[6]   Rodoni, B.C., Dale, J.L. & Harding, R.M. (1999). Characterization and expression of the coat protein-coding region of the banana bract mosaic potyvirus, development of diagnostic assays and detection of the virus in banana plants from five countries in Southeast Asia. *Archives of Virology* 144: 1725-1737.

[7]   Spencer, P.S., Roy, D.N., Ludolph, A., Hugon, J., Dwivedi, M.P. & Schaumburg, H.H. (1986) Lathyrism: evidence for role of the neuroexcitatory aminoacid BOAA. *Lancet* 2(8515):1066-7.

[8]   Wang, M.B., Abbott, D.C., Upadhyaya, N.M. & Waterhouse, P.M. (2001). Agrobacterium tumefaciens mediated transformation of an elite barley cultivar with virus resistance nd reporter genes. *Australian J Plant Physiology*. 28: 149-156.

[9]   Wesley, S.V., Helliwell, C.A., Smith, N.A., Wang, M.B., Rouse, D.T., Liu, Q., Gooding, P.S., Singh, S.P., Abbott, D., Stoutjesdijk, P.A., Robinson, S.P., Gleave, A.P., Green, A.G. & Waterhouse, P.M.(2001). Construct design for efficient, effective and high throughput gene silencing in plants. *Plant J.* 27: 581-590.

In: Recent Trends in Biotechnology and Microbiology
Editors: Rajarshi Kumar Gaur et al., pp. 117-130

ISBN: 978-1-60876-666-6
2010 Nova Science Publishers, Inc.

Chapter 10

AMPLIFIED FRAGMENT LENGTH POLYMORPHISM (AFLP) AS MOLECULAR MARKER- A PCR BASED TECHNIQUE AND ITS APPLICATIONS IN PLANT SCIENCES

Mangal Singh Rathore, *Saroj Sharma, D. Purohit and R.K. Gaur*

Biotechnology Laboratory, Department of Sciences, FASC, Mody Institute of
Technology and Science (MITS) Deemed University, Lakshmangarh, Sikar
(Rajasthan- 332311) India

ABSTRACT

Detection and analysis of genetic variations help us to understand the molecular basis of various biological phenomena in plants. A molecular marker is a particular segment of deoxyribonucleic acid (DNA) that is representative of the differences at the genome level. As genomes of all plants cannot be sequenced, therefore molecular markers and their correlation to phenotypes provide requisite landmarks for elucidation of genetic variation. Genetic or DNA based marker techniques such as restriction fragment length polymorphism (RFLP), random amplified polymorphic DNA (RAPD), simple sequence repeats (SSR) and amplified fragment length polymorphism (AFLP) are routinely being used in evolutionary, phylogenic and genetic studies of plants. AFLP technique combines the power of RFLP with the flexibility of PCR-based technology by ligating primer recognition sequences (adaptors) to the restricted DNA and selective PCR amplification of restriction fragments using a limited set of primers. The AFLP technique generates fingerprints of any DNA regardless of its source, and without any prior knowledge of DNA sequence. Most AFLP fragments correspond to unique positions on the genome and hence can be exploited as landmarks in genetic and physical mapping. The technique can be used to distinguish closely related individuals at the sub-species level and can also map genes. Significant progress in crop productivity has been made in India due to painstaking efforts of plant breeders; however future possibilities of crop improvement include development of new and more efficient plant ideotypes with

* Email- mangalrathore@gmail.com

improved quality traits. This objective can be successfully achieved, if conventional plant breeding is supplemented with molecular breeding approaches i.e. including both the transgenic crops and the marker-assisted selection (MAS).

INTRODUCTION

Plants are the most beautiful creation of Nature that forms the basis of almost all life on earth, providing protection and sustenance to organisms ranging from bacteria to large mammals. With their unique capacity for photosynthesis, they laid the basic foundation of biological food web, while producing oxygen (O_2) and mopping up excess levels of the greenhouse gas, carbon dioxide (CO_2). They provide not only nutrition but also cater to the secondary needs of man, besides performing a number of important environmental services like recycling of essential nutrients, stabilizing soil, protecting water recruitment areas and helping rainfall control etc. In fact, plants account 93 % of world's food supply and even the remaining 7 % is indirectly contributed by plants through animal products (Vasil, 1990). India (8-30° N and 68-97.5° E), as a sub-continent is one of the world's 12 leading biodiversity centers. The climatic, altitudinal and latitudinal variations, coupled with varied ecological habitats of this country contributed the development of immensely rich vegetation. The sub-continent encompasses 16 different agro-climatic zones, 10 vegetation zones, 25 biotic provinces and about 426 habitats of specific species. Myer *et al.* (2000) included two hot spots (Western Ghat and Eastern Himalayas) from India in their updated list of world's biodiversity hotspots. It has been estimated that the Indian sub-continent contains about 45,000 plant species (20 % of world population). Besides all these facts about richness of biodiversity, it has been widely recognized that loss of genetic diversity is a major threat for the maintenance and adaptive potential of species. For many plant species *ex situ* as well as *in situ* conservation strategies have been developed to safeguard the extant genetic diversity. To manage this genetic diversity effectively the ability to identify genetic variation is indispensable. Characterization of diversity has long been based on morphological traits mainly. However, morphological variation is often found to be restricted and genotype expression may be affected by environmental conditions, thereby constraining the analysis of genetic variation. Therefore, molecular genetic techniques are nowadays being applied as a complementary strategy to traditional approaches in the management of plant genetic resources.

The use of traits in plant as markers for their genetic relationship predates genetics itself. The concept of genetic markers is not a new one; in the 18th century, Carl Linnaeus (later *Carl von Linne'*) used the number and arrangement of plant's sexual organs to determine their systematic relationship. Gregor Mendel (Father of Genetics) derived his principles of inheritance by following visible traits (phenotype-based genetic markers) in the progeny of sexual crosses in the nineteenth century. Later, phenotype based genetic markers for *Drosophila* led to the establishment of the theory of genetic linkage. The limitations of phenotype based genetic markers led to the development of more general and useful direct DNA based markers that became known as molecular markers. A molecular marker is defined as a particular segment of DNA that is representative of the differences at the genome level. Molecular markers should not be considered as normal genes, as they usually do not have any biological effect, and instead can be thought of as constant landmarks in the genome. They are identifiable DNA sequences found at specific locations of the genome and transmitted by the standard laws of inheritance from one generation to the next. They are used to flag the position of a particular gene or the inheritance of a particular characteristic. In a genetic cross, the characteristics of interest will usually stay linked with the molecular markers. Thus, individuals can be selected in which the molecular marker is present, since the marker indicates the presence of the desired characteristic. Molecular markers rely on a DNA assay, in contrast to morphological markers, based on visible traits, and biochemical markers, based on proteins produced by genes. Molecular markers may or may not correlate with phenotypic expression of a trait. Molecular markers offer numerous advantages over conventional phenotype based alternatives as they are stable and detectable in all tissues regardless of growth, differentiation, development, or defense status of the cell are not confounded by the environment, pleiotropic and epistatic effects.

WHAT ARE GENETIC MARKERS?

Genetic markers represent genetic differences between individual organisms or species. Generally, they do not represent the target genes themselves but act as 'signs' or 'flags'. Genetic markers that are located in close proximity to genes (i.e. tightly linked) may be referred to as gene 'tags'. Such markers themselves do not affect the phenotype of the trait of interest because they are located only near or 'linked' to genes controlling the trait. All genetic markers occupy specific genomic positions within chromosomes (like genes) called 'loci' (singular 'locus').

There are three types of genetic markers: (1) *morphological* (also 'classical' or 'visible') *markers* which themselves are phenotypic traits or characters; (2) *biochemical markers*, which include allelic variants of enzymes called isozymes; and (3) *DNA* (or *molecular*) *markers*, which reveal sites of variations in DNA (Jones *et al.*, 1997; Winter and Kahl, 1995). Morphological markers are usually visually characterized phenotypic characters such as flower colour, seed shape, growth habits or pigmentation. Isozyme markers are differences in enzymes that are detected by electrophoresis and specific staining. The major disadvantages of morphological and biochemical markers are that they are limited in number and influenced by environmental factors or the developmental stage of the plant (Winter and Kahl, 1995). However, despite these limitations, morphological and biochemical markers have been extremely useful to plant breeders (Eagles *et al.*, 2001; Weeden *et al.*, 1994).

DNA markers are the most widely used type of marker predominantly due to their abundance. They arise from different classes of DNA mutations such as substitution mutations (point mutations), rearrangements (insertions or deletions) or errors in replication of tandemly repeated DNA (Paterson, 1996). These markers are selectively neutral because they are usually located in non-coding regions of DNA. Unlike morphological and biochemical markers, DNA markers are practically unlimited in number and are not affected by environmental factors and/or the developmental stage of the plant (Winter and Kahl, 1995). Apart from the use of DNA markers in the construction of linkage maps, they have numerous applications in plant breeding such as assessing the level of genetic diversity within germplasm and cultivar identity (Baird *et al.*, 1997; Jahufer *et al.*, 2003; Winter and Kahl, 1995).

The publication of Botstein *et al.* (1980) about the construction of genetic maps using restriction fragment length polymorphism (RFLP) was the first reported molecular marker technique in the detection of DNA polymorphism. Basic molecular marker techniques can be classified into three categories based on the method of their detection: (1) *hybridization-based* e.g. RFLP; (2) *polymerase chain reaction* (PCR)-based e.g. Random Amplified Polymorphic DNA (RAPD), Amplified Fragment Length Polymorphism (AFLP) etc and (3) *DNA sequence-based* e.g. Microsatellites, Single Nucleotide Polymorphism (SNPs) etc (Gupta *et al.*, 1999; Joshi *et al.*, 1999). Essentially, DNA markers may reveal genetic differences that can be visualized by using a technique called gel electrophoresis and staining with chemicals (ethidium bromide or silver) or detection with radioactive or colourimetric probes. DNA markers are particularly useful if they reveal differences between individuals of the same or different species. These markers are called polymorphic markers, whereas markers that do not discriminate between genotypes are called monomorphic markers. Polymorphic markers may be codominant or dominant based on whether markers can discriminate between homozygotes and heterozygotes. Codominant markers indicate differences in size whereas dominant markers are either present or absent. Strictly speaking, the different forms of a DNA marker (e.g. different sized bands on gels) are called marker 'alleles'. Codominant markers may have many different alleles whereas a dominant marker only has two alleles.

Desirable Characteristics of an Ideal Marker

An ideal molecular marker technique should have the following criteria:

- Be polymorphic and evenly distributed throughout the genome
- Provide adequate resolution of genetic differences
- Generate multiple, independent and reliable markers
- Simple, quick and inexpensive
- Need small amounts of tissue and DNA samples
- Have linkage to distinct phenotypes and
- Require no prior information about the genome of an organism

Unfortunately no molecular marker technique is ideal for every situation. Techniques differ from each other with respect to important features such as genomic abundance, level of polymorphism detected, locus specificity, reproducibility, technical requirements and cost. Depending on the need, modifications in the techniques have been made, leading to a second generation of advanced molecular markers.

AMPLIFIED FRAGMENT LENGTH POLYMORPHISM (AFLP)

In the mid 1990's, to overcome the limitation of reproducibility associated with RAPD, AFLP technology (Vos *et al.*, 1995) was developed. Unlike other methods (marker techniques), the technique is patented (AFLP^{TM} is a trademark of Keygene, Wageningen, The Netherlands). With this technique, DNA treated with restriction enzymes is amplified with

PCR. It allows selective amplification of restriction fragments giving rise to large numbers of useful markers, which can be located quickly and reliably on the genome. Thus it combines the power of RFLP with the flexibility of PCR-based technology by ligating primer recognition sequences (adaptors) to the restricted DNA and selective PCR amplification of restriction fragments using a limited set of primers (see the Fig. 1 for a schematic representation of technique). The primer pairs used for AFLP usually produce 50–100 bands per assay. Number of amplicons per AFLP assay is a function of the number of selective nucleotides in the AFLP primer combination, the selective nucleotide motif, GC content, and physical genome size and complexity. The AFLP technique is also referred as *Selective Fragment Length Amplification* (SFLA) or *Selective Restriction Fragment Amplification* (SRFA).

The AFLP technique generates fingerprints of any DNA regardless of its source, and without any prior knowledge of DNA sequence. Most AFLP fragments correspond to unique positions on the genome and hence can be exploited as landmarks in genetic and physical mapping. The technique can be used to distinguish closely related individuals at the sub-species level (Althoff *et al.*, 2007) and can also map genes. Applications for AFLP in plant mapping include establishing linkage groups in crosses, saturating regions with markers for gene landing efforts (Yin *et al.*, 1999) and assessing the degree of relatedness or variability among cultivars (Mian *et al.*, 2002). For high-throughput screening approach, fluorescence tagged primers are also used for AFLP analysis. The amplified fragments are detected on denaturing polyacrylamide gels using an automated ALF DNA sequencer with the fragment option (Huang and Sun, 1999).

Analytical Procedures for AFLP Technique

Analytical procedure for AFLP technique can be summarized in following steps (for schematic representation see figure 1):

- Extraction of DNA.
- Digestion of DNA by restriction endonuclease.
- Ligation of oligonucleotide adaptors to the DNA fragments.
- Selective (pre-selective/selective) amplification of part of the DNA fragments.
- Separation of the fragments by polyacrylamide gel-electrophoresis.
- Visualization of the polymorphisms by autoradiography, silver staining or fluorescence.
- Data analysis.

The detailed description of procedure is as follows.

DNA extraction and Restriction digestion: To prepare an AFLP template, genomic DNA is isolated from source and digested with two restriction endonucleases simultaneously. This restriction digestion generates the substrate/DNA fragments for ligation and subsequent amplification. The restriction fragments for amplification are generated by two restriction endonucleases, for example *EcoR I* and *MSe I. EcoR I* has a 6-base pair recognition site, and *MSe I* has a 4-base pair recognition site. When used together, these enzymes generate small

DNA fragments (with *EcoR I / MSe I* or both kinds of ends) that will amplify well and are in the optimal size range (less than ~1 kb) for separation on denaturing polyacrylamide gels. Due to primer design and amplification strategy, these *EcoR I* – Mse 1 fragments are preferentially amplified (rather than *EcoR I* – *EcoR I* or Mse 1 – Mse 1 fragments). The success of the AFLP technique is dependent upon complete restriction digestion; therefore, much care should be taken to isolate high quality genomic DNA, intact without contaminating nucleases or inhibitors.

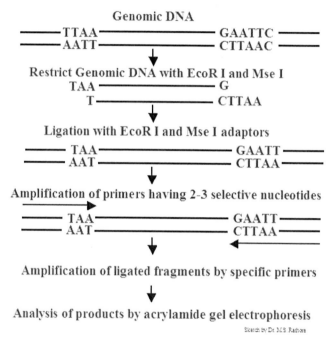

Figure 1. A Schematic representation of AFLP technique: Rare cutting and frequent cutting enzymes restricted Genomic DNA simultaneously, which results in fragments with overhanging ends. The restricted fragments were ligated to end specific adaptors and using primers with random ends (2–3 nucleotides) a set of fragments is amplified and analysed by either agarose and polyacrylamide electrophoresis.

Ligation of adaptors: Adaptors are short double stranded DNA fragments, which have either one or both sticky ends corresponding restriction enzyme. These are used to convert the ends of DNA fragments as per experimental needs. Following heat inactivation of the restriction endonucleases, the genomic DNA fragments are ligated to *EcoR I* and *MSe I* adaptors to generate template DNA for PCR amplification. These common adaptor sequences flanking variable genomic DNA sequences serve as primer binding sites on these restriction fragments. Using this strategy, it is possible to amplify many DNA fragments without having prior sequence knowledge.

Amplification reactions: PCR is performed in two successive reactions. In the first reaction, called pre-amplification, genomic DNAs are amplified with AFLP primers each having one elective nucleotide. The PCR products of the pre-amplification reaction are diluted and used as a template for the selective amplification using two AFLP primers, each containing two-three selective nucleotides. This two-step amplification strategy results in consistently cleaner and more reproducible fingerprints with the added benefit of generating enough template DNA for thousands of AFLP reactions.

The most important factor in determining the number of restriction fragments amplified in a single AFLP reaction is the number of selective nucleotides in the selective primers. Usually the selective primers in several AFLP Analysis Systems contain three selective nucleotides. The second factor in determining the number of amplified fragments is the C and G composition of the selective nucleotides. In general, the more cytosine (Cs) and guanine (Gs) used as selective nucleotides in the amplification primers, the fewer DNA fragments amplified. Also, the smaller the genome being analyzed, the fewer fragments amplified and the simpler the fingerprint.

AFLP reaction: An AFLP reaction usually involves two oligonucleotide primers, one corresponding to the *EcoR I* ends and the other corresponding to the *MSe I* ends. One of two primers are radioctively labeled, preferably the *EcoR I* primer. The primers are end labeled using [γ^{33}p] ATP and T4 polynucleotide kinase. After this, PCR reactions are performed. The PCR conditions differ depending on the nature of the selective extensions of the AFLP primers used for amplification. AFLP reactions are performed using the following cyclic conditions (note that these conditions are generalized):

1) AFLP reactions with primers having none or single selective nucleotides is performed for 20 cycles with DNA denaturation at 94 °C for 30 sec, annealing at 56 °C for 1 min, and extension at 72 °C for 1 min.

2) AFLP reactions with primers having two or three selective nucleotides is performed for 36 cycles with DNA denaturation at 94 °C for 30 sec, annealing for 30 sec, and extension at 72 °C for 1 min. The annealing time in the first cycle is 65 °C, subsequently reduced in each cycle by 0.7 °C for the next 12 cycles, and is continued at 56 °C for the remaining 23 cycles.

AFLP fingerprinting of complex genomes involves amplification in two steps. Pre-amplification, is performed with two AFLP primers having a single selective nucleotide. After pre-amplification, the reaction mixtures are diluted approximately 10-fold, and used as template for second amplification reaction. The second amplification reactions are performed with AFLP primers having longer selective extensions.

Separation of amplified fragments on PAGE and Gel analysis: Products from the selective amplification are separated on a 5% or 6% denaturing polyacrylamide (sequencing) gel. After electrophoresis or separation of amplified fragments, gels are fixed for 30 min in 10% acetic acid dried on the glass plates and exposed to phosphoimage screens for 16 h. Fingerprint patterns are visualized using phosphoimage analysis system. Individual band intensity, size distribution of amplified products, and overall pattern should be similar for AFLP analysis with the same primer pairs and the same DNA template, and will vary between different genomic DNA samples and different primer pairs. Fingerprints of related plants should display common bands, as well as some difference in banding pattern due to DNA polymorphisms. The total number of bands as well as the number of polymorphisms will depend on the crop, variety, and complexity of the genome and primer pair used. Some primer pairs work better in specific crops for identifying polymorphisms, while some primer pair combinations may result in either too few or too many bands for a particular sample. In the case of too few bands, a primer pair containing fewer Gs and Cs is used in the selective nucleotides. In the case of too many bands, a primer pair containing more Gs and Cs is used in the selective nucleotides. The resultant banding pattern ("fingerprint") can be analyzed for polymorphisms either manually or using analytical software (Vos *et al.*, 1995).

Advantages of AFLP Techniques as Molecular Marker

- No sequence data for primer construction is required: It requires no prior knowledge of DNA sequence.
- More reliable and reproducible: It depend less on DNA quality and lab conditions.
- High genomic abundance.
- Number of polymorphic loci that can be detected is 10-100 times greater than RAPD or SSRs.
- Rapid generation and high frequency of identifiable AFLP polymorphism.
- Random distribution throughout the genome, although clustering around centromers has been reported.
- Generation of many informative bands per reaction.
- High reproducibility.
- Wide range of applications.
- Amenable to automation.

Disadvantages of AFLP Techniques as Molecular Marker

- Purified, high molecular weight DNA required.
- Band profiles cannot be interpreted in terms of loci and alleles.
- Dominance of alleles.
- Similar sized fragments may not be homologous.
- Expensive to generate, as detection of bands by silver staining or fluorescent dyes.

INDIAN SCENARIO IN CONTEXT TO MARKER TECHNOLOGY

Significant progress in crop productivity has been made in India during the last 40 years due to painstaking efforts of plant breeders. Some of the most important achievements include introduction of semi-dwarf wheat and rice, hybrid varieties of millets and cotton and noblisation of sugarcane. Future possibilities of crop improvement include development of new and more efficient plant ideotypes with improved quality traits (including nutritional traits), which did not receive attention of plant breeders in the past. This objective can be successfully achieved, if conventional plant breeding is supplemented with molecular breeding approaches including both, the transgenic crops and the marker-assisted selection (MAS).

Table 1.

S. No.	Name of Institution (India)	Markers used	System for experimentation	Area of Research
1.	Agriculture Research Institute, Naini	RFLP	Brassica	Genome and QTL mapping
2.	Center for Cell and Molecular Biology (CCMB), Hyderabad	AFLP	Rice	Diversity study
3.	Ch. Charan Singh University, Meerut	RFLP, SSR, STS, AFLP, SAMPI, EST	Wheat and Barley	DNA fingerprinting, Diversity study, Gene tagging, Genome and QTL mapping and Association analysis
4.	International Centre for Genetic Engineering & Biotechnology (ICGEB), New Delhi	RFLP, RAPD and AFLP	Rice	Genome and QTL mapping, Physical mapping
5.	International Crops Research Institute for the Semi-Arid Tropics (ICRISAT), Patancheru	RFLP. Isozymes, RAPD, AFLP, SSR	Pulses and millets	Germplasm characterization, Diversity study, Genome and QTL mapping, Characterization of cytoplasmic male sterility systems
6.	Jawaharlal Nehru University (JNU), New Delhi	AFLP	Chickpea	Diversity study
7.	M.S. Swaminathan Research Foundation, Chennai	RFLP, RAPD, AFLP	Mangroves and millets	Diversity study
8.	M.S. University, Baroda	RAPD	Rice	Gene tagging
9.	National Botanical Research Institute (NBRI), Luknow	-	Amaranthus	Diversity study
10.	National Chemical Laboratory, (NCL), Pune	RAPD, ISSR, SCAR	Wheat and Chickpea	Diversity study, Gene tagging,
11.	National Research Center for Plant Biotechnology, Indian Agriculture Research Institute (IARI), New Delhi	RAPD, SSR, AFLP, ISSR	Brassica and Rice	DNA fingerprinting, Diversity study, Gene tagging and Genome & QTL mapping,
12.	National Research Center on Plant DNA Fingerprinting, New Delhi	RAPD, SSR, AFLP	All major crops	DNA fingerprinting and Diversity study
13.	Tata Energy Research Institute (TERI), New Delhi	RFLP, RAPD, SSR	Brassica, Poplar, Neem and Withania	DNA Fingerprinting, Germplasm characterization, Diversity study and Genome and QTL mapping
14.	University of Agriculture Sciences, Bangalore	RAPD	Rice	Genome and QTL mapping
15.	University of Delhi, New Delhi	RFLP, RAPD and AFLP	Mustard and *Vigna* sp	DNA fingerprinting, Germplasm characterization and Gene tagging

Although at a number of centers in India, a beginning has been made in the area of MAS (see table 1: the table is taken from Gupta and Roy, 2002), however the progress in the development and use of molecular markers for MAS has been rather slow not only in India, but also at the international level. Considerable work on molecular markers, relevant to crop improvement has been undertaken in India. Different area of molecular marker research being pursued at the different centers in India along with the plant systems and the marker systems that have been used is summarized in Table 1. It will be seen that in India molecular markers in plants are mainly being used for the following purposes:

- Characterization of the available germplasm through DNA fingerprinting and estimation of genetic diversity (sometimes also for testing genetic purity) in this germplasm with the suggested aim of using this information for selection of parents for hybridization programs.
- The tagging of genes/QTL for qualitative and quantitative traits through marker-trait association for MAS.
- The preparation of molecular maps and their use for interval mapping of QTL for polygenic traits.

APPLICATIONS OF MOLECULAR MARKER IN PLANT SCIENCE

Molecular markers are looked as tools for a large number of applications ranging from localization of a gene to improvement of plant varieties by marker-assisted selection. They have also become extremely popular for phylogenetic or taxonomic analysis. The development of molecular markers provides easy, fast and automated assistance to scientists and breeders. Genome analysis based on molecular markers has generated a vast amount of information and a number of databases are being generated to preserve and popularize it. The straightforward applications of the AFLP technique in plant mapping and in marker-assisted breeding include establishing linkage groups in crosses, saturating regions with markers for gene landing efforts (Yin et al., 1999) and assessing the degree of relatedness or variability among cultivars (Mian et al., 2002), variety identification isolation of markers tightly linked to specific genes, and marker assisted backcrossing This section present a brief overview of application of molecular marker in plant science.

Mapping and tagging of genes: Plant improvement either by natural selection or through the efforts of breeders, depends on creating, evaluating and selecting the right combination of alleles. The manipulation of a large number of genes is basically required for improvement of morphological, physio-biochemical traits of the plant. With the use of molecular markers it is now a routine to trace valuable alleles in a segregating population and mapping them. The first genome map in a plant was reported in *Maize* (Helentjaris et al., 1986; Gardener et al., 1993), followed by Rice (McCouch, 1988), *Arabidopsis* (Chang et al., 1988) etc. using RFLP markers. Maps have since then been constructed for several crops (Table 1).

Once mapped, the markers can be efficiently employed in tagging several traits that are important for breeding programs like yield, disease resistance, stress tolerance, seed quality, etc. A large number of *monogenic* and *polygenic loci* for various traits have been identified in a number of plants, which are currently being exploited by breeders and molecular biologists for marker-assisted selection of crops and medicinal plants. The tagged genes can also be used for detecting the presence of useful genes in the new genotypes generated in a hybrid program or by other methods like genetic transformation.

Detection of DNA methylation: In eukaryotic organisms, DNA methylation has been reported as an important phenomenon as it has both epigenetic and mutagenic effects causing differential gene expression, cell differentiation, chromatin inactivation, genomic imprinting and carcinogenesis (Gonzalgo and Jones, 1997). Detection of DNA methylation in tissues of eukaryotic organisms depends on the application of bisulfites or isoschizomers. In general, any method capable of displaying polymorphism of digested DNA fragments can be used to detect DNA methylation (Xiong et al., 1999). AFLP technique acts as a powerful tool for detecting DNA methylation in eukaryotic organisms. Basically for DNA methylation detection, the standard protocol of AFLP fingerprinting is applied with slight modification where MSe I adaptor and primer is substituted with novel Hpa II – Msp I adaptor and primer. Tissue-specific bands are produced only when unique DNA methylation events occur in one of the two genomic DNA samples. This type of methylation-linked AFLP pattern is assumed to be useful as a probe for methylation-related gene(s) as they may correlate with alteration of gene expression in specific tissues (Rossi et al., 1997).

Genotyping of Plants/cultivars: Genotyping is the most reliable method for the identification of lines and varieties. Therefore the DNA fingerprinting methods can be used to analyze the purity of seed lots. Genetic distance analysis can be a powerful tool for breeders to identify different heterotic groups and to increase the efficiency of finding crosses with good combinability. To determine the genetic distance between lines and groups of lines, the lines are fingerprinted and the marker-presence or absence is scored for each line. Based on the obtained score table, similarity indices can be calculated for all combinations of lines. Subsequently, the relatedness amongst the lines can be visualized using a dendrogram. This kind of information is useful for quantification of genetic diversity, characterization of accessions in plant germplasm collections and taxonomic studies.

Diversity analysis of exotic germplasm: After domestication, genetic variation in crops has continued to narrow due to continuous selection pressure for specific traits i.e. yield, thus rendering them more vulnerable to disease and insect epidemics. This reduces the potential for sustained genetic improvement of crops. Thus it is extremely important to study the genetic composition of the germplasm of existing modern-day cultivars in comparison with their ancestors and related species. This will not only provide information on their phylogenetic relationship but will also indicate a chance of finding new and useful genes.

Indirect selection can be an advantageous method of selection in plant breeding. Especially for traits for which the phenotypic tests are unreliable or expensive, markers can offer a solution. Once linked markers have been identified, the AFLP or markers can be converted into simple PCR assays, which allow screening of large numbers of plants for the trait of interest in a cost effective manner. Many DNA markers both specific as well as arbitrary have been used, for DNA fingerprinting of various classes of germplasm. AFLP has also gained popularity as marker for the study of genetic polymorphism especially in species where polymorphism is extremely rare using other types of markers. These kinds of studies help in the classification of existing biodiversity among plants, which can be further, exploited in wild gene introgression programs.

Though there are contradictory reports on this aspect, however, molecular markers are used to test the genetic purity of material or germ plasm stored for conservation.

Phylogenetic and evolutionary analysis: Most of the evolutionary concepts were based on morphological and geographical variations between organisms. It is evident that the techniques from molecular biology provide detailed information about the genetic structure of natural population. RFLP, AFLP, DNA sequencing, and a number of PCR-based markers are being used extensively for reconstructing phylogenies of various species. The taxonomic classification is the first step to determine whether particular germplasm is a part of the primary, secondary or tertiary gene pool of the system concerned. In this context, molecular markers are used for studying the genetic variation in plants and living beings, so as to understand their evolution from wild progenitors and to classify them into appropriate groups. Recent studies reveal the utility of RAPD, AFLP and ISSR (inter simple sequence repeats) markers in evolutionary studies of wheat, rice and other crops.

LIMITATIONS, CHALLENGES, NEEDS AND OPPORTUNITIES

Molecular marker research for crop improvement has its own limitations. Although in several crops, molecular markers have been used for estimations of genetic diversity, none of these studies could be utilized for selecting parents for the hybridization programs for crop improvement. The use of molecular markers for testing genetic purity of germplasm after long-term storage for conservation also has its own limitations. Similarly, molecular markers have been developed for important traits in several crops, but their use for MAS in actual plant breeding is not as visible as earlier anticipated. Globally, the only examples where MAS has been successfully used in practical plant breeding include breeding for resistance against soybean cyst nematode (SCN) in USA, pyramiding of genes for resistance against bacterial blight in rice at International Rice Research Institute (IRRI), Philippines in collaboration with Punjab Agriculture University (PAU), Ludhiana (India) (Singh et al., 2001) and breeding for resistance against gall midge in rice undertaken by International Centre for Genetic Engineering Biotechnology, New Delhi (ICGEB) with rice breeding centers elsewhere in the country.

The major challenge and future needs for molecular marker research for crop improvement in India (Gupta and Roy, 2002) include

- Unfortunately, most of the molecular marker research in India has been done by those who are not the practicing plant breeders, so that an awareness among the practicing plant breeders about the utility of MAS in crop improvement needs to be developed.
- The practicing plant breeders need to build infrastructure in terms of physical facilities, and the skill/expertise in terms of man power.
- In each crop, the plant breeders need to identify the traits for which the cost of MAS can be justified. Once these traits are known, molecular markers may be developed exclusively for these traits.
- Validation studies for the molecular markers already developed need to be undertaken using near-isogenic lines or other breeding material with known genetic constitutions.
- Work need to be started in India on single nucleotide polymorphisms (SNPs) sooner than later, since these will be the markers of choice in future. Molecular marker research in India needs to catch up with the developments at the international level.

- Bioinformatics tools need to be used for making effective use of massive databases that are available, which is rather expensive.

REFERENCES

[1] Althoff, D.M., Gitzendanner, M.A. & K. A. Segraves. (2007). The utility of amplified fragment length polymorphisms in phylogenetics: a comparison of homology within and between genomes. *Syst. Biol.* 56:477-484.

[2] Baird, V., Abbott, A., Ballard, R., Sosinski, B. & S. Rajapakse. (1997). DNA Diagnostics in Horticulture, p. 111-130 *In: P. Gresshoff (Ed.), Current Topics in Plant Molecular Biology:* Technology Transfer of Plant Biotechnology. CRC Press, Boca Raton.

[3] Botstein, D., White, R.L., Skolnick, M. & R.W. Davis. (1980). Construction of a genetic linkage map in man using restriction fragment length polymorphisms. *Am. J. Hum. Genet.* 32:314-333.

[4] Chang, C., Bowman, A.W., Lander, E.S. & E.W. Meyerowitz. (1988). Proc. Natl. Acad. Sci. USA 85:6856-6860.

[5] Eagles, H., Bariana, H., Ogbonnaya, F., Rebetzke, G., Hollamby, G., Henry, R., Henschke, P. & M. Carter. (2001). Implementation of markers in Australian wheat breeding. *Aust. J. Agric. Res.* 52:1349-1356.

[6] Gardiner, A, Kikuch, F. & H. Hirochika. (1993). Jap. J. Genet, 68: 195-204.

[7] Gonzalgo, M.L. & P.A. Jones. (1997). Rapid quantification of methylation differences at specific sites using methylation-sensitive single primer extension (Ms-SNuPt). *Nucleic Acid Res* 25: 2529-2531.

[8] Gupta, P., Varshney, R., Sharma, P. & B. Ramesh. (1999). Molecular markers and their applications in wheat breeding. *Plant Breed.* 118: 369-390.

[9] Gupta P.K. & J.K. Roy (2002). Molecular markers in crop improvement: Present status and future needs in India. *Plant Cell, Tissue and Organ Culture* 70: 229-234.

[10] Helentjaris T., Slocun M., Wright S., Schaefer A. & J. Nienhuis (1986). *Theor. Appl. Genet.* 72: 761-769.

[11] Huang J. & M. Sun (1999). A modified AFLP with fluorescence labelled primers and automated DNA sequencer detection for efficient fingerprinting analysis in plants. *Biotechnol. Techn.* 14:277-278.

[12] Jahufer M., Barret B., Griffiths A. & D. Woodfield (2003). DNA fingerprinting and genetic relationships among white clover cultivars. In: J. Morton (Ed.), *Proceedings of the NewZealand Grassland Association,* Vol. 65, pp. 163-169, Taieri Print Limited, Dunedin.

[13] Jones N., Ougham H. & H. Thomas (1997). Markers and mapping: We are all geneticists now. *New Phytol* 137: 165-177.

[14] Joshi S., Ranjekar P. & V. Gupta (1999). Molecular markers in plant genome analysis. *Curr Sci* 77: 230-240.

[15] McCough S.R., Kochert G.,Yu Z.H., Wang Z.Y., Khush G.S., Cottonan W.R. & S.D. Tanksley (1988). *Theor. Appl.Genet.* 76:815-829.

[16] Mian M.A.R., Hopkins A.A. & J.C. Zwonitzer (2002). Determination of genetic diversity in tall fescue with AFLP markers. *Crop Sci* 42:944-950.

[17] Myers N., Mittermeier R.A., Mittermeier C.G., daFonseca G.A.B. & J. Kent (2000). Biodiversity hotspots for conservation priorities. *Nature* 403: 853-858.

[18] Paterson A.H. (1996). Genome mapping in plants. Academic Press, Inc. and RG Landes Co, Newyork and Austin pp. 330

[19] Rossi V., Motto M. & L. Pelligrini (1997). Analysis of methylation pattern of maize. Opaque-2 (O_2) promoter and in vitro binding studies indicate that the O_2 β – Zip protein and other endosperm factor can find to methylated target sequence. *J. Biol. Chem.* 272:13758-13765.

[20] Vasil I.K. (1990). The realities and challenges of plant biotechnology. *BioTechnology* 8:296-301.

[21] Vos P., Hogers R., Bleeker M., Reijans M., van de Lee T., Hornes M., Frijters A., Pot J., Peleman J., Kuiper M. & M. Zabeau (1995). AFLP: a new technique for DNA fingerprinting. *Nucleic Acids Res* 23:4407-4414.

[22] Weeden N., Timmerman G. & J. Lu (1994). Identifying and mapping genes of economic significance. *Euphytica* 73: 191-198.

[23] Winter P. & G. Kahl (1995). Molecular marker technologies for plant improvement. *World Journal of Microbiology & Biotechnology* 11: 438-448.

[24] Xiong, L.Z., Xu, C.G., Saghai, Maroof, M.A. & Q.F. Zhang. (1999). Pattern of cytosine methylation in an elite rice hybrid and its parental line by a methylation sensitive amplification polymorphism technique. *Mol. Gen. Genet.* 261:439-446.

[25] Yin, X., Stam, P., Dourleijn, C.J. & M.J. Kropff. (1999). AFLP mapping of quantitative trait loci for yield-determining physiological characters in spring barley. *Theor. Appl. Genet.* 99: 244-253.

In: Recent Trends in Biotechnology and Microbiology
Editors: Rajarshi Kumar Gaur et al., pp. 131-138

ISBN: 978-1-60876-666-6
2010 Nova Science Publishers, Inc.

Chapter 11

Mycorrhiza and its Significance in Crop Improvement

Harish Dhingra and Sakshi Issar

Department of Science, Mody Institute of Technology and Science, Laxmangarh, Distt. Sikar, (Rajasthan)

Abstract

Mycorrhiza means when fungi enter into a mutualistic relationship with the plant roots. In this type of relationship, fungi actually become integrated into the physical structure of the roots. This sort of association helps both the partners i.e. fungus and its host plant to be mutually benefited by each other as the fungus absorbs water and nutrients from soil and supplies the same to the plant and in turn derives its nutrition (carbohydrates and photosynthates) from the host plant for its growth and multiplication.

Introduction

Mycorrhizal associations exist for prolonged periods with the maintenance of a healthy physiological interaction between plant and the fungus. Enhanced uptake of water & mineral nutrients, particularly phosphorus & nitrogen, has been noted in many mycorrhizal associations; plants with mycorrhizal fungi are therefore able to occupy habitat they otherwise could not. The mycorrhiza is of three types i.e. (i.) ectotrophic mycorrhiza (ii) endotrophic mycorrhiza and (iii) ecto-endotrophic mycorrhiza (Jeffries & Barea, 1994).

Ectotrophic Mycorrhiza

Ectotrophic Mycorrhiza or ectomycorrhizae are common in gymnosperm and angiosperm, including most oak, beech, birch, and coniferous trees. Ectomycorrhizal fungi generally have the optimal growth temperature of 15-30°C and are acidophilic with the optimum growth at pH 4-6 or as low as 3 (Auge, 2001). This is an association of the fungus and the feeder roots (root hairs) in which the fungus grows predominantly intercellularly in the cortical region penetrating the epidermis by secreting proteolytic enzymes and develops extensively outside the root forming a network of hyphae which is called 'hurting net' or the 'fungus mantle'. The mantle is of variable thickness, colour and texture surrounding the rootlets (Zak & Parkinson, 1982). Ectomycorrhiza absorbs and stores plant nutrients like nitrogen, phosphorus, potassium and calcium etc. in their mantle. Besides it also converts some organic molecules into simple, easily available forms. In the infected roots the hyphae radiate from the mantle into the soil. It changes the morphology of the root system with repeated dichotomous branching and elongation of the ectomycorrhizal fungus (Olsson *et al.* 1998). The infected roots get brightly coloured depending on the colour of the fungal symbionts. The fungus mantle shields the feeder roots from the soil borne pathogens. They are also known to produce some growth promoting substances like cytokinins. This type of mycorrhizal association is commonly found in some forest trees belonging to the families Fragaceae, Butalaceae, Salicaceae etc. The common genera of the plant species include *Eucalyptus, Papulus, Salix, Cedrix, Pinus* and many others. Most of the ectomycorrhizal fungi come in the class basidiomycetes and the common genera are *Amanita, Fuscoboletinus, Lecimum, Boletus, Cortinarious, Suillus, Pisolithus* and *Rhizopogan* etc. Some ectomycorrhizal fungi are capable of producing enzymes such as cellulase, but such activity is normally suppressed within the host plant and therefore, the fungi do not digest the plant roots (Orlowska *et al.* 2002).

ENDOTROPHIC MYCORRHIZA

This is an association between fungus and roots of a plant in which the fungal hyphae infects the roots and remain up to the cortical region (parenchyma of roots) by secreting cellulolytic enzymes. It grows predominantly intracellular i.e. within the root cells (Pearson & Jakobsen, 1993). The hyphae form coils and swellings that eventually disappear as a result of digestion of the invaded cells and the hypha follows the advancing meristematic root tips while its mycelium extends far away in soil and thus increases the surface area for absorption of nutrients. It also occurs in many forest plants including *Taxus, Podocarpus, Cupresus* and *Araucaria* etc. The fungi forming endomycorrhiza mostly belong to the class zygomycetes and family endogonaceae. The important genera of such fungi include *Endogone, Gigaspora, Acculospora, Glomus* and *Sderocystis* etc. Most of the endomycorrhiza commonly occurring in various economically important plants are characterized as vascular-arbuscularmycorrhiza (VAM) recognized by the presence of 'vesicles' (terminal spherical structures containing oil droplets) and 'arbuscles' (complex structures formed by repeated dichotomous branching of hyphae in cortical cells of the feeder roots).

The mycorrhizal mycelium appears to be more resistant than the root itself to abiotic stresses such as drought, metal toxicity, and soil acidity. The fungi increase plant growth through improved uptake of nutrients, especially phosphorus, made possible by the exploration by the external hyphae of the soil beyond the root hair and phosphorus depletion zones (Rillig & Steinberg, 2002, Steinberg & Rillig, 2003). The VAM mycorrhizal association results in increased phosphate uptake by the plant and improved uptake of the other ions, such as zinc, sulfate and ammonium from the soil. The beneficial effect of the mycorrhizae in plant growth is prominent in phosphorus deficient soil. Because of the generally low availability of P in tropical soils, it is having great significance.

ECTOTROPHIC CUM ENDOTROPHIC MYCORRHIZA

This is an association of the fungus and the roots of a plant representing a condition where typical ectotrophic intercellular infection is accompanied with intracellular penetration of hyphae. They are found sometimes in the root system of beech, lodge pole pine and pondersa pine. This sort of mycorrhizal association is considered to be transitional between ectotrophic and endotrophic forms, where infection is typically ectotrophic (intercellular) alongwith endotrophic penetration of hyphae. They are mostly found to occur on the root system of many horticultural crops and tropical trees (Zhou, 1999).

Benefits from Mycorrhiza

Mycorrhizal fungi offer wide range of benefits to the crop plants as their host. They are known to increase the solubility of minerals in soil, improve nutrient uptake of host plants and protect the roots against soil borne phytopathogens with their antagonistic effects. Such beneficial qualities of the mycorrhiza can be well utilized for better crop stand, establishment of high yielding forests, land reclamation and introduction exotic plant species (Harrier, 2001, Hildebrandt & Bothe, 1999, Li *et al.* 1991). Mycorrhiza has greater applicability in enhancing plant growth under tough environmental conditions. Some of the major benefits rendered by mycorrhizal association to the associated plants are briefly discussed below.

Nutrient Uptake by Plants

Experimental findings suggest that mycorrhizal association with forest plants, vegetables and field crops improves the absorption of almost all the nutrients required by them for their growth such as phosphorus, copper, zinc, sulphur, magnesium, manganese and iron etc. Evidences show that mycorrhizal plants in nutrient deficient soils absorb larger amount of nutrients than the non-mycorrhizal ones. During the mycorrhizal association, formation of root hairs are suppressed & fungal hyphae overtake their function (Gaur & Adholeya, 2004). Therefore this association greatly increase the radius of nutrient availability for the plant because of longevity of feeder roots. Nitrogen containing compounds and calcium have also been found to be absorbed into the fungal mycelial sheath, followed by transfer to the plant roots.

Solublization of Plant Nutrients

More root exudates are secreted by the mycorrhizal roots system in the rhizosphere resulting in enhanced activity of useful rhizospheric microbes such as phosphate solublizing microorganisms, organic matter decomposers and symbiotic as well as nonsymbiotic biological nitrogen fixers etc. The fungal hyphae associated with mycorrhizal plants can ramify root system of a plant in the rhizosphere over a large soil volume and provide a greater surface area for nutrient absorption than the roots of a non- mycorrhizal plant.

Stress Tolerance

There are evidences that the plants with mycorrhiza are more tolerant to stress such as soil salinity, alkalinity, acidity and drought conditions. Moreover, by exploitation of larger soil volume, extended root growth and increased absorptive area, the mycorrhizal plants exhibit better growth than the nonmycorrhizal ones especially in the arid and semi arid regions where low moisture and high temperature are very critical for survival and growth of the plants (Tullio *et al.* 2003, Weissenhorn & Berthelin, 1995). Mycorhizal plants are also more tolerant to toxic heavy metals than the non- mycorrhizal plants.

Utilization of Fixed Phosphates and Insoluble Phosphates

Recent advances on mycorrhizal research suggest that the symbiotic mycorrhizal association can lead to more economical use of phosphate fertilizers and better exploitation of cheaper and less soluble rock phosphates. The better utilization of sparingly soluble rock phosphate is explained by the hyphae making closer physical contact than the roots with the ions dissociating at the particle surface.

PROTECTION OF PLANTS FROM ATTACK OF PATHOGENS

In some cases mycorrhiza is found to offer adequate protection to the root system from the attack of pathogenic fungi. For instance, the mycorrhizal fungi such as *Lectarious delicious* and *Boletus* sp. Antagonize *Rhizoctonia solani, Lectarious camphorates, Lectarious* and *Cortinarious* sp. have been found to produce antibiotics known as 'chloromycorrhiza' and 'Mycorrhizin A' which are antifungal to the phytopathogens like *Rhizoctonia solani, Pythium debarynum* and *Fusarium oxysporum* etc. Besides the fungal mantle in the mycorrhizal roots also offers physical resistance to various soil borne pathogenic fungi if the mycorrhizal fungus gains entry into root system prior to infection of root by the potential pathogen (Liao *et al.* 2003). In pine seedlings, the fungus mantle has been found to restrict the penetration of *Phytopthora cinamuni.* Moreover some insoluble polysaccharides are known to accumulate in the cell wall and lignin production is enhanced in the mycorrhizal roots. In such tissues the growth of pathogens like *Fusarium oxysporum* and *Pyrenochaeta terrestires* etc have been found to be considerably restricted. It has now been well established that inoculation of mycorrhizal fungi particularly the ectomycorrhizal ones can protect the roots from soil borne pathogens and nematodes by forming a sheath around the roots and stimulate the plant growth by reducing the severity of diseases.

Production of Growth Hormones

Studies have revealed that plants with mycorrhiza exhibit higher content of growth regulators like cytokinins and auxins as compared to the non-mycorrhizal ones (Laheurte & Oberwinkler, 1990).

Uptake of Heavy Metals

Mycorrhizal association is known to reduce the concentrations of various heavy metals like Zn, Pb, Cd etc (Barea *et at.* 1997, Dehn & Schuepp, 1989, Diaz *et al.*1996). Therefore, they can act as a tool for phytoremediation of heavy metals from the contaminated soil (Jamal *et al.* 2002, Joner, & Leyval, 2000).

CONCLUSION

Pre-inoculation of seeds with suitable mycorrhizal fungi can enhance the growth of the crop plant. Methodology has been developed for preparation and use of vascular-arbuscularmycorrhiza for seed treatment in form of slurry application, soil treatment and seedling treatment. In view of the significant contribution of mycorrhiza to enhance plant growth, there is need of further researches on development and utilization of more potential fungal strains capable of mycorrhizal association and standardization of protocol for their utilization with economically important agricultural and horticultural crops as well as in plantation of forest trees.

Advantages of Mycorrhizal Bio-fertilizer Inoculation:

1) It leads to saving of 20-40 kg of inorganic phosphates per hectare. One ton of VAM is equivalent to 24 tons of phosphorus with the application dose of 0.5 kg/ha spore suspensions.
2) Solubilize and absorb phosphate and sulphur and increase availability and uptake efficiency of plants for secondary and micronutrients, which are relatively insoluble and immobile.
3) Provide plant nutrients like phosphorus, potassium, calcium magnesium sulphur, iron, manganese, zinc and copper etc at a very low cost.
4) Enhances plant growth by release of vitamins and hormones and plant growth substances like auxins and cytokinins etc. Increase in crop yield has been recorded by about 20-40% with their use.
5) Controls soil borne microbial pathogens and nematodes.
6) Improves the physical, chemical and biological properties of soil by organic matter decomposition and soil aggregation.

7) Helps survival and proliferation of beneficial microorganisms like phosphorus solublizers, organic matter decomposers and nitrogen fixers etc.

8) It has no harmful residual effect on soil fertility and plant growth. It sustains productivity as an eco-friendly input.

9) It is required in very small amount and also becomes available to the subsequent crop. Helps in nutrient recycling.

10) Hastens seed germination, flowering and maturity with increased production.

FUTURE PROSPECTS

It is having great future prospects with respect to low cared areas & also in various parts of Rajasthan. However, further researches are required for an understanding of the physiology and ecology of the association and host specificity (which cultivars mostly likely to be benefited from which strain of mycorrhiza) under different agro-climatic conditions. It needs screening and selection of the most efficient fungal endophytes. Development of suitable protocol for production of inoculants on large scale, simplified techniques for their utilization in the field and assessment of the field situations where mycorrhizal use would be most beneficial need to be studied. Suitable protocol needs to be standardized to derive utmost benefit from the dual culture of VAM fungi and biological nitrogen fixers together. Compatibility of mycorrhiza to pesticides, high level of inorganic fertilizers and poor aeration conditions need to be thoroughly investigated.

REFERENCES

[1]   Auge, R. M. (2001) Water relations, drought and vesicular-arbuscular mycorrhizal symbiosis. *Mycorrhiza*, 11: 3–42.

[2]   Barea, J. M., Azco'n-Aguilar, C. & Azcon, R. (1997). Interactions between mycorrhizal fungi and rhizosphere microorganism within the context of sustainable soil–plant systems. In Multitrophic Interactions in Terrestrial Systems (eds Gange, A. C. and Brown, V. K.), Cambridge, United Kingdom, pp. 65–77.

[3]   Dehn, B. & Schuepp, H. (1989). Influence of VA mycorrhizae on the uptake and distribution of heavy metals in plants. *Agric. Ecosyst. Environ.*, 29: 79–83.

[4]   Diaz, G., Azcón-Aguilar, C. & Honrubia, M. (1996). Influence of arbuscular mycorrhiza on heavy metal (Zn and Pb) uptake and growth of Lygedum spartum and Anthyllis cytisoides. *Plant Soil*, 180: 241–249.

[5]   Gaur A. & Adholeya A. (2004). Prospects of arbuscular mycorrhizal fungi in phytoremediation of heavy metal contaminated soils. *Current Science*. 86 (4): 528-534.

[6]   Harrier L. A. (2001). The zinc violet and its colonization by arbuscular mycorrhizal fungi. *Journal of Experimental Botany*. 52: 469-478.

[7]   Hildebrandt, U. & Bothe, H. (1999). The zinc violet and its colonization by arbuscular mycorrhizal fungi. *J. Plant Physiol*. 154: 709–717.

[8] Jamal, A., Visser, P. & Ernst, W. H. O. (2002). Vesicular-arbuscular mycorrhizae decrease zinc toxicity to grasses growing in zinc polluted soil. *Int. J. Phytoremed.* 4: 205–221.

[9] Jeffries, P. & Barea, J. M. (1994). Biogeochemical cycling and arbuscular mycorrhizas in the sustainability of plant soil systems. In Impact of Arbuscular Mycorrhizas on Sustainable Agriculture and Natural Ecosystems (eds Gianinazzi, S. and Schuepp, H.), Birkhauser, Basel, pp. 101–115.

[10] Joner, E. J. & Leyval, C. (2000). Uptake of Cd by roots and hyphae of a Glomus mosseae/ Trifolium subterraneum mycorrhiza from soil amended with high and low concentrations of cadmium. *Biotechnol. Lett.*, 22: 1705–1708.

[11] Laheurte, F. & Oberwinkler, F. (1990). Element localization in mycorrhizal roots of Pteridium aquilinum L. Kuhn collected from experimental plots treated with cadmium dust. *Symbiosis* 9: 111–116.

[12] Li, X. L., George, E. & Marschner, H. (1991). Phosphorus depletion and pH decrease at the root-soil and hyphae-soil interfaces of VA mycorrhizal white clover fertilized with ammonium. *New Phytol.* 119: 397–404.

[13] Liao, J. P., Lin, X. G., Cao, Z. H., Shi, Y. Q. & Wong, M. H. (2003). Interactions between arbuscular mycorrhizae and heavy metals under sand culture experiment. *Chemosphere*, 50: 847–853.

[14] Olsson, P. A. George, E. & Marschner, H. (1998). Contribution of an arbuscular mycorrhizal fungus to the uptake of cadmium and nickel in bean and maize plants. *Plant Soil*, 201: 9–16.

[15] Orlowska, E. Zubek, S., Jurkiewicz, A., Szarek-ukaszewska, G. & Turnau, K. (2002). Influence of restoration on arbuscular mycorrhiza of *Biscutella laevigata* L. (Brassicaceae) and *Plantago lanceolata* L. (Plantaginaceae) from calamine spoil mounds. *Mycorrhiza*, 12: 153–160.

[16] Pearson, J. N. & Jakobsen, I. (1993). The relative contribution of hyphae and roots to phosphorus uptake by arbuscular mycorrhizal plants, measured by dual labelling with 32P and 33P. *New Phytol.* 124: 489–494.

[17] Rillig, M. C. & Steinberg, P. D. (2002). Glomalin production by an arbuscular mycorrhizal fungus: a mechanism of habitat modification. *Soil Biol. Biochem.*, 34: 1371–1374.

[18] Steinberg, P. D. & Rillig, M. C. (2003). Differential decomposition of arbuscular mycorrhizal fungal hyphae and glomalin. *Soil Biol. Biochem.*, 35: 191–194.

[19] Tullio, M. Pierandrei, F., Salerno, A. & Rea, E. (2003). Tolerance to cadmium of vesicular arbuscular mycorrhizae spores isolated from a cadmium-polluted and unpolluted soil. *Biol. Fertil. Soils* 37: 211–214.

[20] Turnau, K. Kottke, I. & Oberwinkler, F. (1993). Element localization in mycorrhizal roots of Pteridium aquilinum L. Kuhn collected from experimental plots treated with cadmium dust. *New Phytol.*, 123: 313–324.

[21] Weissenhorn, I. & Berthelin, J. (1995). Bioavailability of heavy metals and abundance of arbuscular mycorrhiza in a soil polluted by atmospheric deposition from a smelter. *Biol. Fertil. Soil*, 19: 22–28.

[22] Zak, J. C. & Parkinson, D. (1982). Analysis of the mycorrhizal potential in the rhizosphere of representative plant species from desertification-threatened Mediterranean shrublands. *Can. J. Bot.*, 60: 2241–2248.

[23] Zhou, J. L. (1999). Zn biosorption by Rhizopus arrhizus and other fungi. *Appl. Microbiol. Biotechnol.*, 51: 686–693.

In: Recent Trends in Biotechnology and Microbiology
Editors: Rajarshi Kumar Gaur et al., pp. 139-154

ISBN: 978-1-60876-666-6
2010 Nova Science Publishers, Inc.

Chapter 12

BIOTECHNOLOGY OF POST-HARVEST DISEASES OF FRUITS AND VEGETABLES

Sudhir Chandra[1], Raghvendra P. Narayan[2] and H. K. Kehri[1]

Department of Botany, University of Allahabad, Allahabad 211002[1]
MITS, University, Lakshmangarh, Sikar 332311[2]

ABSTRACT

India has a variety of soils and climates. Therefore, almost all kinds of fruits and vegetables, *viz.*, temperate, sub-tropical and tropical are being grown in different agroclimatic regions of the country. This natural advantage promises great potential for the country's fruit and vegetable culture. Development of good fruit and vegetable industry on sound scientific basis could be of benefit in many ways. They enrich human diet by supplementing vitamins, minerals and sugars in addition to being an easily digestible food.

The importance of fruits and vegetables for our country, which is facing acute shortage of food can neither be denied nor disparaged. It is unfortunate that country suffers a great loss due to considerable damage caused to fruits and vegetables by a number of diseases occurring during post-harvest period (transit and storage). These are highly perishable things and the losses are more considerable than as often realized, because fruits and vegetables increase manifold in unit value while passing from the field at harvest to the consumer. There are more than 250 known parasitic diseases of fruits and vegetables that cause decay and blemishes during transit, marketing and storage. The damage and losses incurred vary with the crop, growing conditions in the field, handling during post-harvest and transit, and storage conditions. In a developed country like U.S.A., where advanced post-harvest technology is applied, annual loss of fruits and vegetables is approximately to the tune of 200 million dollars. According to an old estimate New York city alone suffers losses of 700 car loads of fruits and vegetables every year. In India exact data on losses are not available, however, the data collected from some past studies put the average loss of fruits and vegetables at 20-30 percent (Bose *et al.*, 1993).

INTRODUCTION

Post-harvest losses in fruits and vegetables may be categorised as losses due to physical and physiological factors, and those due to pathogenic microorganisms. Mechanical injuries include all cuts, bruises, punctures, abrasions, insect scars, hailscars, crushing, cracking and freezing. The parasitic diseases, decay and other defects caused by fungi, bacteria and insects constitute a well defined category while non-parasitic disorders include the various physiological responses of fruits and vegetables to the post-harvest environment.

These diseases not only reduce the market value of the fruits and vegetables but also affect the economy of the industries engaged in preparation of fruit and vegetable products. The present communication deals with the advances made in this field during past 50 years in our country.

The post-harvest diseases of fruits and vegetables have been studied from time to time in different countries. Some of the important contributions in this field have been made by Baker (1938), Rose et al., (1951), Ramsey and Smith (1961), Purse (1953), Oxenham (1961), Kieley and Long (1960), Harvey and Pentzer (1960), Chorin and Rotem (1961), Geard (1961), Beraha (1962), Roth (1963), Matthee and Ginsberg (1963), Greene and Goos (1963), Smith et al. (1964), Simmond (1965), Eckert and Sommer (1967) and Eckert (1977).

Initially the plant pathologists of our country were concentrating on study of diseases of fruits and vegetables under field conditions. Later on, however, attention has been paid on the diseases occurring during post-harvest period and attempts are still being made to control the diseases in order to reduce the losses. So far there has been little detailed work on such diseases in our country. The work is scattered and generally symptoms of various diseases, their causal organisms and occasionally methods of their control were mentioned. Some of the noteworthy contributions include the work of Mitter and Tandon (1929), Mehta (1939), Chona (1933), Ghatak (1938), Singh (1943), Sinha (1946), Bhargava and Gupta (1957), Damodaran and Ramakrishnan (1963), Tandon and Verma (1964), Srivastava et al. (1964a, 1964b, 1965), Rao (1966), Rao and Bhide (1984), Chandra and Khanna (1996).

Dastur (1916) while working on rot of bananas recommended the removal and destruction of infected fruits. Dey and Nigam (1933) noted a large number of apples rotting due to soft rot. They studied the disease caused by *Aspergillus niger* and recommended the wrapping of the fruits in tissue paper before packaging. Chona (1933) studied the effect of temperature and humidity on the incidence of stem end rot of banana caused by species of *Botryodiplodia* and *Gloeosporium*. Kheswalla (1936) carried out investigations on fruit diseases and described the symptoms of blue mould of apple (*Penicillium expansum* Link) and pink rot of apple (*Trichothecium roseum* Link). For the control of such diseases he recommended careful handling of the fruits to avoid bruises or injury to the skin. Mehta (1939) studied the effect of temperature and pH on the growth of *Rhizopus arrhizus* Fisher, which causes apples to rot. Tandon (1950) showed that ripe guavas were also susceptible to *Pestalotia* isolated from apples. Grewal (1954) carried out investigations and studied in detail the effect of temperature, humidity, light, age of the fruits and variety on the incidence of disease. Srivastava et al., (1964b) observed that some of the fungal pathogens causing the spoilage of fruits in transit and storage were also found associated with other parts of the plants in orchards. On the basis of his detailed studies Tandon (1970) highlighted some of the problems associated with post-harvest diseases of fruits and vegetables. Thakur et al. (1974)

advocated the use of growth regulators in the control of post-harvest diseases of fruits and vegetables. Rao and Bhide (1984) on the basis of his extensive survey of market and storage diseases of fruits and vegetables in Bombay and Maharashtra, highlighted some ecopathological problems. Chandra (1986) advocated the use of homoeopathic drugs in the control of post-harvest microbial spoilage of fruits and categorized the method to be safe and economical.

Realizing the importance of losses of fruits and vegetables during post-harvest period, the plant pathologists of the country are presently concentrating on new control measures which are warranted in view of the technological and environmental changes. This is justified from the fact that Committee of International Society of Plant Pathology on Chemical Control (Anonymous, 1980) has also emphasized on the search for new control agents.

POST-HARVEST SPOILAGE AND THE CAUSES

Non-Pathogenic Causes

When a large number of fruits and vegetables are bulked together, most of them get rotted. This is not directly due to the pathogens; e.g. when potatoes are stored in a poorly ventilated room, the level of CO_2 may become excessively high in the atmosphere, leading to a corresponding decrease in the O_2 contents. In potato this results in a condition known as black heart. Externally the tubers look normal, but when they are cut open, the central portion appears black. Black heart is caused by lack of oxygen which kills the tissues. It is favoured by the high temperatures - about 32°C - and poor ventilation during transit. Apples suffer from brown heart in a similar fashion. Various types of discolourations and scalds may appear on the skins of various fruits as a result of poor storage conditions. Factors responsible for bringing about discolouration are thought to be volatile substances, probably esters. In the case of apple scald, a hypothesis is that it is due to the precursors of volatile substances, to materials of low volatility or to toxic substances accumulating beneath the cuticle.

Pathogenic Causes

Apart from non-pathogenic causes a large number of fungi and bacteria are also responsible for storage diseases of fruits and vegetables. Species of *Rhizopus* cause considerable loss to peaches, grapes, strawberries, sweet potatoes, cucurbits, crucifers, tomatoes and egg plants. *Rhizopus* spp. produce soft rot of the fleshy parts which proceeds rapidly at high temperatures. There is often leakage of juices from the affected parts of fruits and vegetables. Under humid conditions, the typical cottony, coarse and stringy mycelium of *Rhizopus*, with characteristic black sporangia, covers the diseased tissue. In temperate parts, stone and pome fruits are affected by brown rot fungi as *Sclerotinia fructigina, S. laxa* and *S. fruiticola*. The fruits get shriveled and mummified and become a mass of dried fruit tissue completely penetrated by the mycelium of the fungus and covered with conidia.

Vegetables, such as carrots, potatoes, onions and celery are attacked by soft rot bacteria. The soft rot is brought about by the action of enzymes (pectolytic and cellulolytic) on the pectic substances and cellulose of the cell walls. As a result of these enzymes the host tissues disintegrate and their protoplasts die. *Erwinia* spp. *viz., E. carotovora, E. atroseptica, E. aroidae* and *E. chrysanthemi*, are mainly responsible for bacterial rots in storage. The factors which predispose the fruits and vegetables to soft rots under stored conditions are humidity, temperature, presence of bruises or lesions or attack by other pathogens. Some other bacteria also cause soft rot at high temperatures and high humidity.

MODE OF INFECTION AND PATHOGENESIS

Harvested fruits and vegetables are vulnerable to attacks by microorganisms because of their high moisture content and rich nutrients. During harvesting, packaging and transportation injuries of various kinds are caused which facilitate the entry of certain pathogens. The injury caused during severing the fruits and vegetables from the plants is a frequent point of initiations of post-harvest diseases.

Physiological injuries caused by low or high temperatures, deficiency of oxygen and other such factors make the fruits and vegetables prone to attacks by the pathogens. Storage at chilling temperatures increases their susceptibility to infections, e.g. stem end rot of grapes and Nigrospora rot of bananas. High temperatures also make them liable to severe storage rots during post-harvest period. Injuries due to hot-water treatment to eliminate the infection of certain pathogens also make them susceptible to certain other pathogens e.g. citrus fruits become more prone to decay due to *Penicillium* when the fruits are given hot water treatment to eliminate infection of *Phytophthora* sp.

In many post-harvest diseases of fruits and vegetables, the pathogen responsible for the spoilage enters them before harvesting. For example, spores of *Colletotrichum gloeosporoides* germinate in moisture on the surface of banana, mango and papaya fruits during their development on the tree. The infection remains latent and produces the disease during post-harvest period. Other examples of latent infection include *Diploidia* and *Phomopsis* sp. infection in citrus fruits causing stem end rot, *Phytophthora* sp. infection in citrus causing brown rot in storage. Nearly all the pathogens responsible for diseases in storage initiate the disease in the same manner. After gaining entry they all produce extracellular enzymes, e.g. pectolytic and cellulolytic enzymes and toxins which start the degenerative process in advance of the fungal hyphae or bacterial cells of the attacking pathogens. Majority of them are saprophytes and may obtain nourishment from dead or dying tissue. However, they are very selective in their nutritional requirements when they attack any host. They influence the storage substances by absorbing them after converting some of the complex forms into simpler ones. The market as well as nutritive value of the fruits and vegetables is thus lowered either by ugly appearance or due to change in the stored products of the hosts.

Various metabolites of the host tissues suffer modifications in their composition due to infection by a pathogen and hence the nutritional value of the stored products is lowered. In storage diseases of fruits and vegetables, changes in metabolites have been widely studied. Alterations in the levels of various carbohydrates, vitamins, organic acids, amino acids, etc. have been investigated by workers.

Nitrogenous compounds are important storage products in fruits and the protein metabolism is closely related to other metabolic activities. Infection of the fruits or vegetables causes many changes in their amino acid contents. Absence or decrease in the concentration of amino acids in infected fruits or vegetables may possibly be attributed to their preferential utilization by the fungus or to their degradation by enzymes or their simultaneous utilization in synthesis of proteins.

Organic acids are important constituents of the fruits and vegetables. Changes in organic acids in the fruits during pathogenesis has been reported by a number of workers. The disappearance or decrease in organic acids has generally been attributed to their utilization by the pathogen. Increase in concentration or appearance of new organic acids during pathogenesis may be assigned to an interaction between the host and pathogen.

A number of investigators have reported marked qualitative and quantitative differences in the sugar contents in healthy and diseased fruits and vegetables. The decrease in carbohydrate built up may be due to following :

(i)   Break down of carbohydrates by the enzymes produced by the pathogens,

(ii) Increased respiration of the infected tissue,

(iii) Utilization of carbohydrates as substrate for the production of metabolites.

FACTORS INFLUENCING INFECTION AND SPREAD

The ways fruits and vegetables are harvested are important because many pathogens involved in storage diseases enter through the injuries caused during harvesting. Careless handling during marketing, transportation and storage also makes them prone to infection by various pathogens. A number of pathogens are known to initiate the disease during post-harvest period through the injuries caused by insects, e.g. injuries to citrus fruits caused by fruit fly makes them susceptible to attack by fruit-rotting pathogens such as *Penicillium digitatum* and *Penicillium italicum.*

Propagules of pathogens are abundant in the atmosphere and on the surface of fruits and vegetables as they approach maturity in the orchards/fields. The inoculum for the infection during their storage may be received either directly from the atmosphere or from the containers used for their transportation or storage, water used for cleaning fruits and hands or tools of the labourer.

Symptoms of many of the storage diseases are expressed during post-harvest period but actual cause of the disease remains present as latent infection from the time the fruits and vegetables are in the fields (Rao and Bhide, 1984). Initiation of latent infection depends on the level of inoculum available and the environmental conditions to which the orchards or fields are exposed. Thus, in such cases the post-harvest diseases are often influenced by the location of the orchards and fields and the manner in which the fruit trees and vegetable plants are raised.

Fruits and vegetables continue to respire even after their harvest. The resultant heat accelerates the respiration and aging which in turn makes them susceptible to the attack by the pathogens. The rate of disease development very much depends on the temperature, humidity and aeration during transportation and storage. Fruits and vegetables which have a high rate of respiration are more prone to storage diseases. Transportation and storage at low temperature is essential for such fruits and vegetables. On the other hand, for the fruits and vegetables which have a low respiration rate, high temperature during transportation and storage is essential because of the risk of damage to them due to low temperature.

Physiological injuries caused by cold, heat, oxygen deficiency and other environmental agents predispose fresh agricultural products to post-harvest diseases.

HOST-SUSCEPTIBILITY AND MULTIPLE INFECTIVITY

Most fruits and vegetables are susceptible to a given fungus when in contact with the stored produce, under a special type of storage environment. This phenomenon of "host-susceptibility" is correlated to the distribution of the fungus in nature in general and at the "phases" of possible lapses through which the host product is transported after harvest. The phases during which storage fungi get successfully associated with the susceptible hosts include harvest, post-harvest, transport, storage and marketing. The infection at harvest phase is mostly attributable to those pathogens which are carried from the field, such as: anthracnose disease of brinjal caused by *Colletotrichum capsici; C. lindemuthianum* in *Vigna sinensis* and *Phaseolus vulgaris*, besides some fusarial rots of potatoes. In majority of cases, field fungi pose comparatively less pathological problem, e.g., cigar-end disease of banana (in the latent stage) which develops in storage due to *Deightoniella torulosa* in combination with *Colletotrichum gloeosporioides*, rot of papaya; *Pestalotia* rot (*P. psidii*) of guava and custard apple.

As regards infection after harvest, there exists a certain definite extent of host-fungus specificity among the admixture of such fungi usually referred to as 'storage fungi'. Examples of such host-fungus associations include *Rhizopus atrocarpi* on jack fruit, *Trichothecium roseum* and *Cladosporium herbarum* on papaya, *Thielaviopsis paradoxa*, on pineapple, *Thielaviopsis paradoxa* on muskmelon, *Alternaria solani* on tomato, *Fusarium solani* on potato.

In addition to these associations, there are examples of fungi infecting more than one plant product in storage. This phenomenon of multiple infectivity is highly specific and sometimes concurrent, upon one and the same host, giving distinguishable symptoms and colonies of more than one fungus.

The fungi which are commonly associated and well known as 'saprophytes' exist in storage with an obligate relationship with fruits and vegetables. Conversely, well known pathogens also exist in storage on fruits and vegetables. A microbe which coexists in association with a given plant product following harvest and that can be detected under such constant associations, even in widely separated agroclimatic localities is referred as pathogen under storage conditions. The intensity and prevalence of such associations, however, are markedly influenced by meteorological factors. Therefore, incidence of one disease of a given product in one locality may not assume same proportion at some other locality. It may be mentioned that the storage fungi mainly behave like the saprobes rather than true pathogens. In other words, it would be erroneous to consider such storage fungi as true pathogens except those which occur in standing crops under the field conditions aptly called 'field diseases', which are distinct from storage diseases.

POST-HARVEST DISEASES OF FRUITS AND VEGETABLES IN INDIA

Fruits

Mango, guava and banana are the important tropical fruits of the country. The choicest fruit, mango is well known over the world. In India it is planted in about 2 million acres of land, that is 60% of the total area under fruits. Innumerable varieties of mango are known in the country and nearly all of them suffer great losses during post-harvest period. Species belonging to *Rhizopus, Aspergillus, Colletotrichum, Botryodiplodia, Pestalotia, Phomopsis* and *Diplodia* have been reported to cause rotting of fruits during marketing, transit and storage.

Guava is another common fruit in the country. A number of fungi are known to attack the fruits in storage. Species of *Aspergillus, Botryodiplodia, Pestalotia, Phoma, Phomopsis* and *Rhizopus* have been reported to cause rotting of fruits. *Gloeosporium* sp. not only cause anthracnose but rotting as well.

Banana another important tropical fruit suffers from many diseases during post-harvest period. *Colletotrichum musae, Fusarium roseum, Verticillium theobromae, Ceratocystis paradoxa* have been reported to cause crown rot. *Colletotrichum musae (Gloeosporium musarum)* has been reported to cause cigar-end disease as well as anthracnose.

Papaya is cultivated all over the tropical and subtropical regions of the world. In India, it is cultivated in a number of states including Tamil Nadu, Maharashtra, Bengal, Bihar, Uttar Pradesh and Madhya Pradesh. During post-harvest period the fruits are known to suffer due to a number of rots caused by forms belonging to *Botryodiplodia, Phytophthora, Ascochyta, Trichothecium, Cladosporium*. Species of *Colletotrichum* and *Gloeosporium* have been reported to cause rotting as well as anthracnose during storage.

Citrus fruits (*viz.* lemon, musambi, orange) are one of the richest sources of vitamin C. The fruits are highly prone to fungal decay during transit, marketing and storage. A number of fungi belonging to the genera *Phomopsis, Diplodia, Alternaria, Sclerotium, Penicillium* and *Ceratocystis* have been shown to be responsible for the rotting of the fruits. The fruits are also susceptible to *Colletotrichum. Trichoderma lignorum* and *Aspergillus aculiatus* have been reported to cause storage rot of mandarin orange. Four species of *Phytophthora viz., P.citrophthora, P.palmivora, P. nicotianae* and *P. arecae* have also been reported to be associated with post-harvest rotting of citrus fruits.

Grape is a sub-tropical fruit and has an additional importance in being a diet for patients. Because of delicacy it is prone to injuries and microbial spoilage during marketing, storage and transit. A variety of fungal forms including species of *Penicillium, Botryodiplodia, Cladosporium, Gloeosporium, Alternaria, Botryotis, Stemphylium* and *Aspergillus* have been reported to cause spoilage of fruits.

Litchi, pineapple, pomegranate and sapota are some of the other tropical fruits, but all of these have a restricted distribution in the country. The total acreage of litchi plantation in this country is about 25, 000 acres, the main bulk of this area lies in Bihar. On an average, losses ranging from 15-20% have been reported due to spoilage of fruits during storage, marketing and transit. A number of fungi including species of *Aspergillus, Colletotrichum* and *Pestalotia* have been reported to cause rotting of fruits.

Sapota, pomegranate and pineapple are known to suffer from losses during storage, however, only few diseases have been reported. *Botryodiplodia theobromae, Pestalotia sapotae* and *Hendersonula toruloidea* are responsible for causing decay of sapota fruits in storage. *Phomopsis versoniana* also causes a severe fruit rot of pomegranate. *Gloeosporium, Fusarium, Penicillium* and *Aspergillus* are other well known organisms causing fruit rots of pomegranate. Probably because of hard covering pineapple fruits are not prone to injuries. *Ceratocystis paradoxa* and *Rhizopus stolonifer* have been shown to be responsible for decay of fruits during storage, transportation and marketing.

Amongst temperate fruits apple and pear occupy an important place in the country. Both of them have restricted cultivation, they are, therefore, transported to long distances. Further, apple being an important fruit for the patients it is kept under storage sometimes for long periods so as to keep the supply maintained in the markets. A variety of fungal forms including *Penicillium, Botrytis, Aspergillus, Rhizopus, Trichothecium, Phoma* and *Gloeosporium* are known to cause rotting of apple as well as pear fruits. In addition, *Cryptosporiopsis* and *Phylactaena* spp. have been reported to induce lenticel rot during post-harvest period. In recent past, few species have been added to the list of species responsible for spoilage of fruits of apple, e.g., *Clathridium corticola, Gliocladium roseum.*

Peach and cherry are the temperate fruits of lesser importance. The area under these fruits is mainly located in Himalayas at various elevations. One of the common rot of these fruits during post-harvest period is caused by *Monilinia fruticola.* In addition, peaches have been shown to be attacked by few other wound pathogens, e.g. *Aspergillus niger* and *Botryodiplodia theobromae.*

Vegetables

Like fruits, vegetables are also prone to microbiological spoilage caused by fungi, bacteria, yeasts and moulds. A significant portion of losses of vegetables during post-harvest is attributed to diseases caused by fungi and bacteria. Succulent nature of the vegetables make them easily invadable by the organisms. Many serious post-harvest diseases of fresh vegetables occur rapidly and cause extensive breakdown of the commodity-sometimes spoiling the entire package. It is estimated that 36% of the vegetable decay is caused by soft-rot bacteria. Obviously the source of infection is soil in the field, water used for cleaning, and surface contact with equipment and storage environment. The most common pathogens causing rots in vegetables are fungi such as *Alternaria, Botrytis, Diplodia, Monilinia, Phomopsis, Penicillium, Rhizopus,* and *Fusarium* and bacteria *Erwinia,* and *Pseudomonas.* High temperature and relative humidity favour the development of post-harvest decay organisms.

Potato is an important starchy crop in both sub-tropical and temperate regions. Even in tropical region it is widely grown during winter season. India ranks fourth in area and fifth in production in the world. However, a number of fungi have been reported to cause the post-harvest spoilage of potato e.g. *Alternaria alternata, Aspergillus fumigatus, Fusarium coeruleum, F. oxysporum, F. trichothecioides, Geotrichum candidum, Sclerotium rolfsii, Trichothecium roseum.* Remarkable decay has been reported by soft rot pathogen *Erwinia carotovora,* dry rot pathogen *Fusarium* spp., Gangrene pathogen *Phoma exigua* var. *foveata,* skin spot pathogen *Oospora pustulans* and silver scurf pathogen, *Helminthosporium solani* during storage.

Tomato is another most popular and widely grown vegetable in the country and is well adapted in all the regions. Being very soft skinned it is attacked by a number of storage fungi and bacteria and cause a considerable damage. Maximum number of pathogens associated to it has been reported by the scientists. These are *Alternaria alternata, A. solani, Aspergillus niger, Cylindrocarpon tonkinense, Cladosporium fulvum, C. tenuissimum, Colletotrichum phomoides, Curvularia lunata, C. lycopersici, Geotrichum candidum, Gilbertella persicariae, Fusarium moniliforme, F. roseum, F. oxysporum* f.sp.lycopersici, *Myrothecium carmichaelii, M. roridum, Nigrospora oryzae, Phoma destructiva, P. exigua, Phytophthora nicotianae var. parasitica, Rhizoctonia solani, Stemphylium vesicarium, Syncephalastrum racemosum, Trichothecium roseum, Cylindrocladium scoparium.* However, Alternaria rot caused by *Alternaria tenuis*, anthracnose by *Colletotrichum* spp. canker by *Corynebacterium michiganense*, soft rot by *Erwinia carotovora*, bacterial spot by *Xanthomonas vesicatoria*, Cladosporium rot by *Cladosporium herbarum*, early blight rot by *Alternaria solani*, Fusarium rot by *Fusarium* spp., Ghost spot by *Botrytis cinerea*, Helminthosporium rot by *Helminthosporium* spp., late blight rot by *Phytophthora infestans*, Phoma rot by *Phoma destructiva* are major post-harvest diseases of tomato.

Cucurbits form an important and a big group of vegetable crops cultivated extensively in this country. This group consists of a wide range of vegetables, either used as salad, or for cooking, or for pickling or as desert fruits or candied or preserved. A number of pathogens have been reported to cause damage to the gouards (*Luffa acutangula, L. cylindrica, Momordica charantia, Lagenaria siceraria, Benincasa cerfera*) e.g. *Alternaria alternata, Botryodiplodia theobromae, Colletotrichum lagenarianum, Fusarium acuminatum, F. moniliforme, F. oxysporum, F. roseum, F. equiseti, Glomerella cingulata, Macrophomina phaseolina, Pythium aphanidermatum,* and *Rhizopus nigricans*; likewise *Drechslera halodes* and *Sclerotium rolfsii* on pumpkin (*Cucurbita moschata*); *Colletotrichum lagenarium, Fusarium roseum, Myrothecium roridum,* and *Pythium aphanidermatum* on cucumber (*Cucumis sativus*); *Colletotrichum capsici, Fusarium equiseti, F. oxysporum, Geotrichum candidum* and *Phoma cucurbitacearum* on tinda (*Citrullus vulgaris*).

Peas and beans are also very common nutritious vegetables grown throughout India. *Pisum sativum, Phaseolus* spp., *Cymopsis tetragonoloba, Vigna catjung, Dolichos lablab* etc. are some of them which suffer heavy losses because of the post-harvest pathogens. *Alternaria alternata, Curvularia lunata, Coleophoma empetri, Colletotrichum capsici, C. lindemuthianum , C pisi, Fusarium roseum, Geotrichum candidum, Phoma cucurbitacearum, P. vignae, Diplodia phaseolina* have been reported to be associated with the rot of these vegetables.

Chillies are the green or dried ripe fruits of *Capsicum annuum* and sometimes *Capsicum frutescens*. It forms an indispensable adjunct in every house in the tropical world. India is one of the major chilli growing country. It is grown in both tropical and subtropical areas of the country. When transported and stored a great number of pathogens attack to spoil them. *Alternaria alternata, Botryodiplodia theobromae, Colletotrichum capsici, Curvularia lunata, Fusarium roseum, Phoma capsici, Phytophthora parasitica, Rhizoctonia bataticola, Rhizopus nigricans, Fusarium solani, F. diversisporum, Cladosporium oxysporum* have been reported to cause a number of post-harvest diseases.

Leafy vegetables are also important in India and are generally cooked with potato, tomato and brinjal. Cabbage, Cauliflower, spinach etc. are some of them. Being very soft and delicate they are easily attacked by the fungi and bacteria causing their decay in transit and market. *Erwinia carotovora* causes the bacterial soft rot of cabbage and *Xanthomonas campestris* is reported to cause the black rot of it. Cauliflower is generally attacked and decayed by fungi. *Alternaria* and *Stemphylium* cause the *Alternaria* rot while *Alternaria brassicae* the brown rot.

CONTROL MEASURES

A. Pre-Harvest Treatments

To avoid the possibility of infection before harvest, spray treatments are given to the fruits in orchards. In case of certain post-harvest diseases such as Rhizopus rot of peaches, brown rot, stem-end rot and Penicillium rot of oranges, spraying the fruits with chemicals such as dichloran (2,6, dichloro 4-nitroaniline), benomyl, thiabendazole, before harvest has been found effective in checking the disease during the post-harvest period. Similar treatments with fungicides and antibiotics have been found successful in anthracnose of mango, banana and papaya. Spray of aureofungin on fruits in orchards has been shown to control a number of post-harvest diseases of fruits in the country.

B. Post-Harvest Treatments

Hot Water Treatment

Since long the treatment of fruits with hot water at a temperature slightly higher than the thermal death point of the suspected pathogen has been one of the convenient methods to eradicate the latent infection.

Low Temperature and Humidity

Low temperature and humidity have a retarding effect on the activities of microorganisms. Hence, storage at low temperature accompanied with low relative humidity is practiced to save the fruits and vegetables from post-harvest microbial spoilage. Due care is taken to see that the storage conditions have no harmful effects on the stored material.

Wrappers

Use of wrappers has been advocated as a protective measure for fruit from post-harvest spoilage. Tandon (1967) evaluated the efficacy of different types of wrappers including polythelene, butter and newspaper and found them effective in improving the keeping qualities of mango and guava fruits. In case of papaya fruits, polythelene wrappers had a negative effect but the use of newspaper wrappers and baskets containing paddy straw gave positive results and decreased the incidence of disease.

Plastic film wrappings over inoculated fruits have been reported to reduce the metabolic activities of fruit rotting fungi and providing protection to the post-harvest decay of mango fruits for long periods.

Radiation

Partial success achieved so far, in ioning radiation as a therapeutic agent in the control of plant diseases, particularly for those which cause serious damages during the post-harvest periods, has focussed the attention of many plant pathologists. However, no attention has been paid to expedite this method in India. Gupta and Chatrath (1973) have been successful in checking the rots of guava fruits by treating the fruits, 5, 18 and 24 hrs after inoculation with gamma radiation at a dose of 100 kr.

Chemical Treatment

A variety of chemicals including fungistats, fungicides, antibiotics and growth regulating substances have been evaluated for their efficacy against various post-harvest diseases of fruits and vegetables.

A variety of systemic and non-systemic fungicides have been recommended for use against different post-harvest diseases of fruits and vegetables.

Non-Systemic Fungicides

Among non-systemic ones, zerlate, captan, dithane M-45, ziram, difoltan 80-W, ferbam and bordeaux mixture have been tried extensively. Ferbam at 3000 ppm have been reported to check *Geotrichum (G. candidum)* rot of Kagzi nimboo. Ferbam was found to be best therapeutant while ziram and captan the best protectants for *Macrophomina (M. phaseolina)* rot of papaya by Kapur and Chohan (1974). Ziram (0.25% - 1%) has been reported to be effective also against *Colletotrichum (C. capsici)* rot of chilli. Similar results have been obtained for other diseases including *Phomopsis (P. destructum)* rot of guava and cottony leak *(Pythium aphanidermatum)* of cucurbit fruits. Calcium Chloride and Ziram has been reported to control to control post harvest decay of pears by Sugar *et al.* (2003).

Systemic Fungicides

Systemic fungicides have given excellent performance and have received sufficient recommendation for a number of diseases.

Antibiotics

Amongst antibiotics, aureofungin has been tried on a number of diseases. Success has been achieved in controlling *Diplodia (D. natalensis)* rot of mango and *Alternaria (A. alternata)* rot of tomato with the help of aureofungin. It has been found to be effective also against *Colletotrichum (C. capsici)* rot of chilli, *Pestalotia (P. psidii)* canker of ripe guava, *Fusarium (F. solani)* and *Rhizopus (R. stolonifer)* rot of papaya, *Aspergillus (A. flavus, A. niger, A. quadrilineatus, A. nidulans, A. variecolor)* rot of litchi. It has been reported to give protection to tomato fruits against few pathogenic fungi *viz., Fusarium moniliforme, Aspergillus niger* and *Curvularia lunata.*

Growth Regulating Substances

Thakur *et al.* (1974) employed growth regulating substances for the control of *Rhizopus* rot of some fruits and vegetables. The soft rot of apple and pomegranate fruits as well as potato tubers caused by *R. arrhizus* were successfully controlled with 2,4-D and 2,4,5-T. These substances were equally effective for brinjal fruit rot incited by *R. stolonifer* and could delay the rotting for 7 days. They were, however, less effective in case of banana soft rot caused by *R. oryzae.*

Other Approaches

Homoeopathic Drugs

Khanna and Chandra (1976a, 1977c, 1978, 1987, 1990) have attempted to explore the possibilities of using homoeopathic drugs for the control of post-harvest spoilage of fruits. Certain potencies of selected drugs not only inhibited the spore germination and growth of pathogenic fungi but also gave protection to fruits against those fungi (Khanna and Chandra, 1995). Guava fruits treated with Kali iodatum 87 and Arsenicum album 181 could be saved from the attack of *Pestalotia psidii* (Khanna and Chandra, 1977c). Kali bichromicum 200 and Lycopodium clavatum 30 were effective against *Botryodiplodia theobromae* Pat. and *Geotrichum candidum* Link ex Pers. responsible for the guava fruit rot. Kali iodatum 148 and Thuja occidentalis 87 were found effective for *Fusarium (F. roseum)* rot of tomato (Khanna and Chandra, 1976a). Similarly Lycopodium 149 was found effective for mango fruits against *Pestalotia mangiferae* (Khanna and Chandra, 1978). Kehri and Chandra (1986) reported Arsenicum album to be effective as a protectant against Botryodiplodia rot of guava fruit (*Botryodiplodia theobromae*). Attempts have also been made to investigate the effect of these drugs on the metabolism of some fungal pathogens. Khanna and Chandra (1992a,b) reported that a number of drugs check the rate of respiration and glucose uptake of spores during germination. It has also been reported that certain drugs alter the permeability of spore membrane during germination (Khanna and Chandra, 1992a, b).

Plant Extracts

Sinha and Saxena (1990a) reported that flower extract of *Lantana camara* suppressed the fruit rot of tomato caused by *Aspergillus niger* and also adversely affected the insect *Drosophila busckii* which is responsible for aggravating the fruit rot. They have also reported the satisfactory control with latex of *Euphorbia hirta* (Sinha and Saxena, 1990b).

Biological Control

Biological control is one of most recent and effective way to control the post harvest diseases. Several bacteria and fungi are used to control post harvest and pre harvest diseases. Wolfgang *et al.,* Rachid *et al.* (2009), Jose *et al.* (2008) Jamalizadeh *et al.* (2008) have shown that selected isolates of *Aureobasidium pullulans, Rhodotorula glutinis, Pichia anomala, Bacillus licheniformis* and *Bacillus subtilis* reduced the size and number of diseases and

lesions on wounded apples caused by the postharvest pathogens *Penicillium expansum,*
Botrytis cinerea, Pezicula malicorticis, Gray mold, fire blight causing pathogens etc.
Thangavelu *et al.* (2007) in their paper proved that bioagents can control crown rot pathogen
(*Lasiodiplodia theobromae*) of banana. Similarly Kalogiannis *et al.* (2006) proved that
phyllosphere yeasts can be used to control grey molds of tomato.

EPILOGUE

After Independence, one of the tasks of the country was to develop a viable and
productive agricultural economy leading to self-sufficiency in our food requirements.
However, self-sufficiency in the true sense can be achieved only when each individual in the
country is assured a balanced diet. Fruits and vegetables are the only natural sources of
protective food as they supply nutrients, vitamins and minerals. In a country where the
population is predominantly vegetarian, this can be achieved by increasing the production and
decreasing the post-harvest decay of fruits and vegetables. Losses of fruits and vegetables
occur due to improper harvesting, transportation, storage and distribution. Post-harvest life of
fruits and vegetables governed by water content, respiratory rate, ethylene production,
endogenous plant hormones and exogenous factors such as microbial growth, temperature,
relative humidity and atmospheric composition. Fruits and vegetables can be conserved by
careful manipulation of these factors. The losses can be minimized by adopting necessary
cultural operations, careful handling and packaging. The use of appropriate chemical at pre-
and post-harvest storage may extend availability of fruits and vegetables over a long period.
The fruits and vegetables can also be effectively stored under controlled and modified
atmosphere to delay senescence and inhibit microbial decay.

REFERENCES

[1]   Anonymous (1980). FAO Plant Prot. Bull., 28:92-106.
[2]   Baker, R.E.D. (1938). Studies in the pathogenicity of tropical fungi II. The occurrence
        of latent infection in developing fruits. *Ann. Bot.* 2:919-931.
[3]   Beraha, L. (1962). Pitting disease of banana. *Plant Dis. Reptr.* 46:354-355.
[4]   Bhargava, K.S. and Gupta, S.C. (1957). Market diseases of fruits and vegetables in
        Kumaon. Cottony leak of beans. *Indian Phytopath.* 10:48-49.
[5]   Bose, T.K., Som, M.G. and Kabir, J. (1993). Vegetable Crops, Naya Prokash, Calcutta.
[6]   Chandra, S. (1986). Post-harvest microbial spoilage of fruits. In: *Rev. Trop. Pl. Pathol.*
        2, Today and Tomorrow's Printers & Publisher, New Delhi: 365-188.
[7]   Chandra, S. and Khanna, K.K. (1996). Post harvest diseases of guava in India. In
        :Disease Scenario in crop plants Vol. I Fruits and Vegetables. (Eds. V.P. Agnihotri, Om
        Prakash, Ram Kisun and A.K. Misra), International Books & Periodicals Supply
        Service, Delhi: 71-80.
[8]   Chona, B.L. (1933). Preliminary investigations on the diseases of bananas occurring in
        the Punjab and their method of control. *Indian Agr. Sci.* 3:673-687.

[9]   Chorin, M. and Rotem, J. (1961). Fruit diseases of the cavandish bananas in Israel. *Israel J. Agr. Res.* 11:43-50.

[10]   Damodaran, S. and Ramakrishnan, K. (1963). Anthracnose of banana. J. Madras Univ. 33B:249-279.

[11]   Dastur, J.F. (1916). Spraying for ripe rot of bananas. *Agr. J. India* 11:142.

[12]   Dey, P.K. and Nigam, V.S. (1933). Soft rot of apples. *Ind. J. Agric Sci.* 3: 663-673.

[13]   Eckert, J.W. (1977). Control of post-harvest diseases. In: Antifungal compounds, Vol. I (Eds. M.R. Siegel and H.D. Sisler), Maecel Dekker Inc., New York.

[14]   Eckert, J.W. and Sommer, N.F. (1967). Control of diseases of fruits and vegetables by post-harvest treatment. *Annu. Rev. Phytopath.* 5:391-433.

[15]   Geard, I.D. (1961). Diseases of peas. *Tasmanian J. Agr.* 13:132-143.

[16]   Ghatak, P.N. (1938). Investigation on orange rot in storage. I. Orange rot due to two strains of *Fusarium moniliforme* Sheldon. *Indian Bot. Soc.* 17:141-146.

[17]   Green, G.L. and Goos, R.D. (1963). Fungi associated with crown rot of boxed bananas. *Phytopath.* 53 :271-275.

[18]   Grewal, J.S. (1954). Cultural and Pathological studies of some fungi causing diseases of fruits. D.Phil. Thesis, University of Allahabad, Allahabad.

[19]   Gupta, J.P. and Chatrath, M.S. (1973). Gamma radiation for the control of post-harvest fruit rot of guava (*Psidium guajava*). *Indian Phytopath.* 26:506-509.

[20]   Harvey, J.M. and Pentzer, W.T. (1960). Market diseases of grapes and other small fruits. U.S. Deptt. Agr. Handbook 189:37pp.

[21]   Jamalizadeh, M., Etebarian, H. R., Alizadeh, A. and Aminian, H.(2008). Biological control of gray mold on apple fruits by *Bacillus licheniformis* (EN74-1). *Phytoparasitica*, 36(1): 23-29.

[22]   Jose Granado, Thurig, B., Kieffer, E., Petrini, L., Fliebach, A., Tamm, L., Franco, P. Weibel, and Gabriela S. Wyss (2008). Culturable Fungi of Stored 'Golden Delicious' Apple Fruits: A One-Season Comparison Study of Organic and Integrated Production Systems in Switzerland. *Microbial Ecology*, 56(4): 720-732.

[23]   Kalogiannis, S., Tjamos, S. E., Stergiou, A., Polymnia, P. Antoniou, Basil N. Ziogas, and Eleftherios C. Tjamos (2006). Selection and evaluation of phyllosphere yeasts as biocontrol agents against grey mould of tomato. *European Journal of Plant Pathology* 116(1): 69-76.

[24]   Kapur, S.P. and Chauhan, J.S. (1974). Evaluation of fungicides for the control of fruit rot of papaya by *Macrophomina phaseolina*. *Indian Phytopath.* 27:251-252.

[25]   Khanna, K.K. and Chandra, S. (1976). Control of tomato fruit rot caused by Fusarium roseum with homoeopathic drugs. *Indian Phytopath.* 29: 269-272.

[26]   Khanna, K.K. and Chandra, S. (1977c). Control of guava fruit rot by Pestalotia psidii with homoeopathic drugs. *Plant Dis. Reptr.* 61:362-366.

[27]   Khanna, K.K. and Chandra, S. (1978). A homoeopathic drug controls mango fruit rot caused by *Pestalotia mangiferae*. *Experientia* 34:1167-1168.

[28]   Khanna, K.K. and Chandra, S. (1987). Spore germination, growth and sporulation of certain fruit rot pathogens as affected by homoeopathic drugs. *Proc. Nat. Acad. Sci.,* India 57 :160-169.

[29]   Khanna, R. and Chandra, S. (1990). Homoeopathic drugs in the control of the fruit rots of guava. *Proc. Nat. Acad. Sci.,* India 50:345-348.

[30] Khanna, K.K. and Chandra, S. (1992a). Effect of homoeopathic drugs on respiration of germinating fungal spores. *Indian Phytopath*. 45:348-353.

[31] Khanna, K.K. and Chandra, S. (1992b). Effect of homoeopathic drugs on membrane permeability of germinating fungal spores. *Nat. Acad. Sci. Letters* 15:167-171.

[32] Khanna, K.K. and Chandra, S. (1995). Management of phytopathogenic microbes through homoeopathic drugs. In: Microbes and Man (Eds. Chandra, S., Khanna, K.K. and Kehri, H.K.), Bishen Singh Mahendra Pal Singh Publication, Dehra Dun: 167-182.

[33] Kheswalla, K.F. (1936). A Phoma disease of Asparagus. *Indian J. Agr. Sci.* 6:800-802.

[34] Kieley, T.B. and Long, J.K. (1960). Market diseases of citrus. New S. Wales Dept. Agr. (Publisher) 16pp.

[35] Mathee, F.N. and Ginsberg, L. (1963). The problem of rot in handling of fruit. *The deciduous fruit grower* 13 :69.

[36] Mehta, P.L. (1939). Fruit rot of Apples caused by species of Rhizopus. *Indian J. Agr. Sci.* 9 :711-718.

[37] Mitter, J.H. and Tandon, R.N. (1929). A cultural study of two fungi found in an Indian hill apple. *J. Indian Bot. Soc.* 8 :212-218.

[38] Oxenham, B.L. (1961). Citrus diseases in Queensland. Qd. Dept. Agr. and Stock Div. *Plant Industry Advisory leaflet*. 620 :18pp.

[39] Purse, G.S. (1953). The fruit rots of the custard apple. *Qd. J. Agr. Sci.* 10 :247-265.

[40] Rachid, L., Sebastien, M., Deborah Clercq, M., Najib S., Creemers, P. , and Jijakli, M. H. (2009) Assessment of *Pichia anomala* (strain K) efficacy against blue mould of apples when applied pre- or post-harvest under laboratory conditions and in orchard trials. *European Journal of Plant Pathology* , 123(1): 37-45.

[41] Ramsey, G.B. and Smith, M.A. (1961). Market diseases of cabbage, cauliflower, turnips, cucumbers, melons and related crops. U.S. Dept. Agr., Agr. Handbook 184 :49pp.

[42] Rao, V.G. (1966). An account of the market and storage doseases of vegetables in Bombay-Maharashtra. *Mycopath. et Mycol. Appl.* 28 :165-176.

[43] Rao, V.G. and Bhide, V.P. (1984). Post-harvest diseases of fruits and vegetables in India : Some eco-pathological aspects. (Eds. K.G. Mukerji, V.P. Agnihotri and R.P. Singh), Print House (India), Lucknow :489-607.

[44] Rose, D.H., AcColloch, L.P. and Fisher, D.F. (1951). Market diseases of fruits and vegetables : Apples, Pears and Quinces. U.S. Dept. Agr. Misc. Pub. 168 :72pp.

[45] Roth, G. (1963). Post-harvest decay of litchi fruit. Dept. Agr. Tech. Serv. *Technical communication* 11 :16pp.

[46] Simmond, J.H. (1965). Papaya diseases. Qd. Dept. Primary Industries Div. *Plant Industry Advisory Leaflet* 837 :13pp.

[47] Singh, U.B. (1943). Some diseases of fruits and fruit trees in Kumaon-I. *ICAR Misc. Bull.* 51 :16pp.

[48] Sinha, S. (1946). On decay of certain fruits in storage. *Proc. Indian Acad. Sci.* 24(B) :198-205.

[49] Sinha, P.S. and Saxena, S.K. (1990a). Effect of flower extract of *Lantana camara* on development of fruit rot caused by Aspergillus niger in the presence of *Drosophila busckii. Acta Bot. Indica* 18 :101-103.

[50] Sinha, P.S. and Saxena, S.K. (1990b). Effect of latex of Euphorbia hirta on fruit rot of tomato caused by Aspergillus niger in the presence of Drosophila busckii. Indian Phytopath. 43 :462-464.

[51] Smith, M.A. Ramsey, G.B. and Green, R.J. (1964). Market diseases of fruits and vegetables. U.S. Dept. Agr. Ext. Cir. 523 :19pp.

[52] Srivastava, M.P., Chandra, S. and Tandon, R.N. (1964a). Post-harvest diseases of some fruits and vegetables. Proc. Nat. Acad. Sci., India 34(B) :339-342.

[53] Srivastava, M.P., Tandon, R.N., Bilgrami, K.S. and Ghosh, A.K. (1964b). Studies on fungal diseases of some tropical fruits-I. A list of the fungi isolated from fruits and fruit trees. Phytopath. Z. 50 :250-261.

[54] Srivastava, M.P., Tandon, R.N., Bhargava, S.N. and Ghosh, A.K. (1965). Studies on fungal diseases of some tropical fruits-III. Some post-harvest diseases of mango (Mangifera indica L.) Proc. Nat. Acad. Sci., India 35 (B) :69-75.

[55] Tandon, M.P. (1950). Some Physiological and Pathological studies of some fungi. D.Phil. Thesis, Allahabad University, Allahabad.

[56] Tandon, R.N. (1967). Post-harvest diseases of tropical and sub-tropical fruits. PL 480 Final Technical Report FG : IN-133, University of Allahabad, Allahabad.

[57] Tandon, R.N. (1970). Certain problems of post-harvest diseases of fruits and vegetables. Indian Phytopath. 23 :1-15.

[58] Tandon, R.N., Tandon, M.P. and Jamaluddin (1974). Current Trends in Plant Pathology. Eds. S.P. Raychaudhury and A. Varma. 209-220.

[59] Tandon, M.P., Bhargava, V. and Jamaluddin (1977). Indian Phytopath. 30 :403-404.

[60] Tandon, R.N. and Varma, A. (1964). Some new storage diseases of fruits and vegetables. Curr. Sci. 33 :625-627.

[61] Thakur, D.P., Chenulu, V.V., Kanwar, Z.S. and Kherata, R.B.S. (1974). Growth regulators in control of post-harvest fungal diseases of fruits and vegetables. Indian Phytopath. 27 :532-536.

[62] Thangavelu, R., Sangeetha, G. and Mustaffa, M. M. (2007). Cross-infection potential of crown rot pathogen (Lasiodiplodia theobromae) isolates and their management using potential native bioagents in banana. Australasian Plant Pathology, 36(6): 595.

[63] Verma, G.S. and Kamal, M. (1951). Rot of Mangifera indica L. caused by Aspergillus. Curr. Sci. 20 :68-69.

In: Recent Trends in Biotechnology and Microbiology   ISBN: 978-1-60876-666-6
Editors: Rajarshi Kumar Gaur et al., pp. 155-167   2010 Nova Science Publishers, Inc.

Chapter 13

NEUROAIDS IN INDIAN SCENARIO

Ashish Swarup Verma[1], Anchal Singh[1], Udai Pratap Singh[1] and Premendra Dhar Dwivedi[2]*

[1]Amity Institute of Biotechnology, Amity University Uttar Pradesh, Sector-125, NOIDA-201303 (UP), India.

[2]Food and Dye Toxicology, Indian Institute of Toxicology Research, PO Box-80, MG Marg, Lucknow-226001 (UP), India.

ABSTRACT

HIV infections are always feared, as its infection can only be controlled and can not be cured. It is almost more than 25 years have gone by with extensive researches on HIV; we failed to develop either any vaccine or a suitable therapy to cure HIV.

Undoubtedly, developments of better anti-retroviral drugs as well as improvement in treatment regimens have significantly prolonged survival among HIV patients. These developments have surely helped HIV seropositives to lead a better and longer life; but, it has certain drawbacks, too. The major draw back is increase in incidences of HIV associated dementia (HAD) and other neurological disorders. These can be commonly defined as NeuroAIDS. The worrying part of this whole saga is that NeuroAIDS affects its patients in their prime of life *i.e.,* ~30-40 years of age.

In India, numbers of HIV seropositives are very high (~2.5 millions). In the recent past, due to sincere efforts by the Indian Government, NACO, UNAIDS and various NGOs, now these patients have better access to improved anti-retroviral treatments. These efforts will help to prolong life-expectancy among HIV seropositives, so that they can lead a quality life. Besides this, high pathogenicity of HIV strain prevalent in India may also be a reason for concern on its impact on HAD.

When we put all these facts together, it definitely suggests that soon India will witness an enormous increase in number of NeuroAIDS patients. It is a high time to initiate appropriate measures to control this upcoming health problem in India.

---

*Corresponding author: Prof. Ashish S. Verma, Amity Institute of Biotechnology, Amity University Uttar Pradesh, Sector-125, NOIDA-201303 (UP), India. Ph: (120)-4392757, Mobile: 9868086202, Email: asverma@amity.edu; ashish-gyanpur@hotmail.com

INTRODUCTION

Human Immunodeficiency Virus (HIV) infection is always dealt with fretfulness, because it leads to an incurable disease known as Acquired Immuno-Deficiency Syndrome (AIDS). A full blown AIDS is considered as final chapter of an individual inflicted with this disease. The causative organism for AIDS is HIV and it is of 2 types, HIV-1 and HIV-2. Among these two types, HIV-1 is more prevalent and has more detrimental consequences compared to HIV-2 infection. The focus of this chapter is neurophsychiatric complications of HIV-1 infection and it is referred as HIV throughout this manuscript.

It is almost 30 years by now, when 1st HIV infection was reported among gays from Los Angeles, USA [Gottlieb *et. al.,* 1981]. At that time, it was considered as an unknown disease with certainty of death. At that time, neither we knew name of the disease nor its causative organism. Most of the earlier cases of AIDS were reported among gay community; due to religious and political reasons behind gay behavior, it became a political hot potato in USA and other parts of world. Therefore a lot of effort and money was spent to find out its causative organism as well as cure for this disease. Soon, scientific researches reported HIV as causative organism and finally the disease was named as AIDS. Ignorance about causes of disease and high mortality among AIDS patients, at that time, was the reason for numerous prevailing myths regarding AIDS. All efforts were directed either to control its spread or to cure AIDS, by therapeutic interventions [Verma *et al.,* 2009].

The first successful drug against HIV infection is Azidothymidine (Azt), which is a reverse transcriptase (RT) inhibitor. This drug was able to control HIV infection, although it has certain limitations *e.g.* HIV develops resistance against the drug. Azt is hematotoxic, and its penetration to Central Nervous System (CNS) is low to negligible. To overcome these shortcomings related with Azt, further studies provided us new RT inhibitors. With time, better understanding about HIV replication was evolved, which has given numerous opportunities to scientists for development of new classes of drugs which acts on novel targets in HIV. These drugs are very effective against HIV. At present, we have different classes of anti-HIV drugs like nucleoside reverse transcriptase inhibitors (NRTI), nucleotide reverse transcriptase inhibitors (NtRTI), non-nucleoside reverse transcriptase inhibitors (NNRTI), protease inhibitors and integrase inhibitors. Highly active antiretroviral treatment (HAART), a combination of different anti-retroviral treatments (ART) is more efficient and effective compared to single drug regimen. These advancements in development of new anti-retrovirals helped to reduce morbidity and mortality among HIV seropositives. As a result, nowadays, HIV seropositive can easily lead a quality life up to 20 years post infection [Verma *et al.,* 2009].

HIV directly infects and injures the CNS and Peripheral Nervous System (PNS). Neurosusceptibility of HIV can cause severe neurotoxicity. Neuroinvasion of HIV and neurotoxicty associated with it, may cause various neurocognitive disorders which can be grouped together as NeuroAIDS. Some of the commonly used names are HIV associated dementia (HAD), HIV associated encephalopathy (HIVE), *etc.* These HIV induced neuropsychiatric disorders are wide ranging (Table 1). At this moment, NeuroAIDS leaves so many questions, i) Is HIV itself responsible for NeuroAIDS? ii) Is NeuroAIDS a secondary complication evolved due to long-term ART? iii) Is low level persistent and chronic HIV infection a major contributing factors towards development of NeuroAIDS? or, iv) Is it a combined effect of all these factors? The present understanding of NeuroAIDS and its causes still remains uncertain.

Table 1. NeuroAIDS: Neuropsychiatric Disorders.

Neuropsychiatric Disorders
1. Epilepsy
2. Physical disability
3. Neuropathic pain
4. Addiction
5. Mania
6. Neurocognitive impairment
7. Mood disorders
8. Anxiety
9. Depression
10. Seizures

Ever increasing use of combination anti-retroviral therapy all over the world has undoubtedly reduced the frequency and severity of neuropsychiatric disorders among HIV seropositives. It is probably due to reduction in incidences of opportunistic infections in HIV patients undergoing HAART regimen because of improvement in their immune status [McCombe *et al.*, 2009]. Still, the incidences of neuropsychiatric complications are as high as 50% or more among HIV patients [McArthur *et al.*, 2005]. There is a change in the spectrum of HIV related neuropsychiatric diseases and with times as they are evolving too, therefore, newer assessment of prevalence of NeuroAIDS might not be a real estimate for the burden of neuropsychiatric disease, in general.

NEUROAIDS AND ITS BURDEN

Even though anti-retroviral treatments (ART) have improved clinical conditions among HIV patients, still they have failed to stop a basal levels of HIV replication. At the start of this decade, a new trend of health related problem among HIV patients have emerged which is known as NeuroAIDS. NeuroAIDS is simply a combination of neurological and psychiatric complications commonly observed in AIDS patients. Among AIDS patients there is always a subset of people who exhibit selective vulnerability for neuropsychiatric phenomenon (neuro-susceptibility) in the form of NeuroAIDS. The significance of this new health related problem can be estimated by the fact that almost ~50% of long term HIV survivors show clinical signs and symptoms of NeuroAIDS. This is a matter of grave concern, as mostly these neuropsychiatric disorders are reported at quite an early age (early to late 30s) among HIV seropositives, even though they are under HAART regimen. It is surprising to note that even very young HIV positive (children) undergoing anti-retroviral treatment have been reported to develop NeuroAIDS symptoms. This early onset of neurological complications or NeuroAIDS among HIV seropositives has a major bearing on their health as well as on health care system, too. NeuroAIDS, on one hand increases cost of patient health care, while on the other hand these patients become incapacitated to perform any productive work in their prime of age. If we combine both these factors together, the financial burden caused by NeuroAIDS patients will become astronomical and get worse with ever increasing number of affected patients. These are compelling facts to recognise the importance of NeuroAIDS and its impact to the society and nation. Cognitive impairments are also associated with HIV infections and have substantial impact on its patients. This can have deleterious implications even on day-to-day activities of patients. Therefore, NeuroAIDS has insurmountable impact on families as well as communities. As NeuroAIDS affects both loss of productivity as well as income of diagnosed patient, which ultimately adds up the financial burden to health care system. Not only this, it causes an extra burden for people who act as caretakers. In resource-limited settings like India with such high rates of HIV infection and NeuroAIDS cases, the toll is likely to be devastating.

NEUROEPIDEMIOLOGY

As per UNAIDS estimates, globally there are approximately 33.2 million people living with HIV. The expected range is in between 30.6-36.1 million. These reports suggested that prevalence of HIV infection is similar among adult men and women. About 2.5 million children below the age of 15 are supposed to be living with HIV till 2007, which ranges in between 2.2-2.6 million. As per 2007, estimates there is 16% (6.3 millions cases) drop in HIV cases globally. 70% of this drop is contributed by 6 countries *viz.,* Angola, India, Kenya, Mozambique, Nigeria and Zimbabwe. In India, the main reason behind this drop was the use of an improved methodology for assessment of HIV infections. Still, ~2.5 million HIV seropositives are living in India, which is a very high number for an incurable infection like HIV [UNAIDS Report, 2007].

In totality, UNAIDS report suggests that African sub-continent is the worst affected for HIV and AIDS, because more than $2/3^{rd}$ (world share) of HIV infected people reside in African continent. A detailed analysis suggests that approximately 6800 persons become infected with HIV everyday while 5700 persons die with HIV/AIDS everyday at global level.

When we talk about the numbers of HIV patients, India is ranked next to African countries, although India is one of the worst affected nations among South-East Asian countries. There are about 2.5 million patients living with HIV infection in India, which accounts for 0.36% prevalence as per national adult population. Interestingly, prevalence trends in India vary significantly from one state to another. Majority of HIV infected people live in Andhra Pradesh, Karnataka, Maharashtra and Tamil Nadu. It is amazing to know that HIV epidemic is even confined to certain districts of above-mentioned states. The epidemiological data on national population based survey suggests that prevalence of HIV in southern states is 5 times or higher compared to northern states. The prevalence of HIV in Uttar Pradesh is 0.07%, in Tamil Nadu is 0.4%, in Maharashtra is 0.69%, in Karnataka is 0.97%, and in Andhra Pradesh is 0.73%, while in Manipur it has been reported up to 1.13%. Good news is that prevalence of HIV infection has been consistently declining among pregnant women in all 4 southern states. Majority of HIV transmission in Indian population is through heterosexual contact with the exception of North East region, where needle sharing is the main reason [Dandona *et al.,* 2006a].

There were few studies, which have demonstrated correlation of neurocognitive behaviour among HIV patients in India [Danadona *et al.,* 2006b]. A huge numbers of HIV seropositives and improved medical treatments for them, warrant us for increased prevalence of NeuroAIDS as an upcoming health issue.

NEUROBIOLOGY OF HIV

HIV virus is roughly spherical in structure measuring about 120 nm in diameter. Taxonomically HIV belongs to *Lentivirus* genus and *Retroviridae* family. Being a member of *Retroviridae* family, it contains single-stranded, positive sense RNA along with reverse transcriptase enzyme (RT) enclosed with in an envelope. The genetic information is transmitted via RNA, as they do not contain DNA as genetic material. To overcome this problem retrovirus has a unique enzyme called as reverse transcriptase for the transfer of genetic information. RT reverse transcribes RNA into DNA (Figure 1).

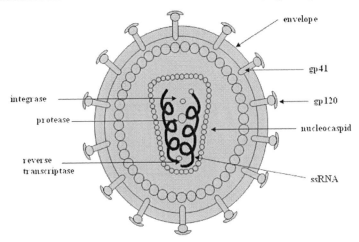

Figure 1: Structure of HIV virion: This is a diagrammatic representation of cross section of HIV virion.

HIV genome size is ~9.8 Kb and consists of 9 genes, which produce 15 different proteins during its replication cycles. These proteins or genes can be divided in 3 categories as per their functions. Gag, Env and Pol are considered as structural proteins, while Tat and Rev are considered as regulatory proteins, whereas Nef, Vpr, Vif and Vpu are accessory proteins (Figure 2).

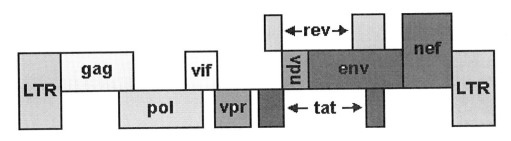

Figure 2: Genetic organization of HIV. Nine genes along with LTR have been depicted in this figure.

Gag is derived from "group-specific-antigen" and its production is regulated by *gag* gene. A 53-kDa precursor protein is synthesized by *gag*, which produces 4 smaller proteins (p7, p9, p17 and p24). Env is derived from "envelope" and encoded by *env* gene. A 160-kDa precursor protein is produced by *env*, which cleaves into gp41 and gp120. Pol is derived from "polymerase" and encoded by *pol* gene. It is a precursor protein which gives rise to 4 different (p10, p32, p51 and p64) proteins.

Tat and Rev are regulatory proteins of HIV as their main function is regulation of viral production. Tat is 110-kDa protein encoded by *tat* gene and this name is derived from "transactivator of transcription". Rev is a 13-kDa protein encoded by *rev* gene and its name has been derived from "regulator of viral expression".

The third groups of genes are specified as "accessory gene", which help in efficient production of new virion and are not essential for replication. These four genes are *nef, vif, vpr* and *vpu*. Nef is a 27-kDa protein encoded by *nef* gene; its name has been derived from "negative factor". Vpr protein is 14-kDa protein encoded by *vpr* gene and its name has been derived from "Viral Protein R". Another gene of this category is *vpu*, which is derived from "Viral protein U". Vif protein is encoded by *vif* gene and its name is derived from "Viral infectivity" [Hult *et al.,* 2008].

NEUROTROPHISM

Initially, cells of lymphocytic and monocytic lineages *viz.*, T-lymphocytes and monocytes were only found to be infected with HIV as they have specific receptors for HIV infectivity. Later, importance of co-receptors like CXCR4 and CCR5 has been recognized for HIV infection. Some of the cells of Central Nervous Systm (CNS) and Peripheral Nervous System (PNS) like astrocytes, microglia, monocyte derived macrophages (MDM), *etc.* express these co-receptors. Occurrence of neuropsychiatric symptoms among AIDS patients lead to further investigation of HIV infectivity in brain/brain cells. The evidences came from autopsy results of AIDS patients' brain. These observations were difficult to explain as brain cells do not express receptors for HIV infectivity. Discovery of co-receptors and their role in HIV infectivity has offered some insight for HIV infectivity to CNS and PNS. Here we will discuss HIV infectivity with reference to CNS only.

CNS has so many cellular components like astrocytes, microglia, MDM, neurons, oligodendrocytes, *etc.* As such, HIV infectivity to nervous system is still debatable, but some types of brain cells express co-receptors for HIV infection, which could be possible reasons behind CNS infectivity. So far, neurons have been non-permissive for HIV infections. Brain cells which get infected with HIV will remain latently infected for longer duration, the trigger for their activation and reactivation is still not well understood.

Astrocytes

Astrocytes are one of the accessory cells of CNS, and their major function is to protect CNS by regulating entry of invader/s to CNS. Astrocytes do not have CD4 receptor, but they express CXCR4 and CCR5 for HIV entry [Gorry *et al.,* 2003]. Several studies have reported HIV infection of astrocytes but mechanism of viral attachment is still unclear. Astrocytes have been demonstrated positive for HIV structural proteins [Brack-Warner *et al.,* 1999] and nucleic acid. Astrocytes remain persistently infected which could end up into productive infection under the influence of inflammatory cytokines [Wang *et al.,* 2004].

Macrophages (Perivascular) and Microglia

Perivascular macrophages and microglia are cells of CNS which come into direct contact with infected cells in perivascular region of CNS. These are resident immunocompetent cells of brain and they respond to any insults to brain. Migratory monocytes reach to CNS due to breach in blood brain barrier (BBB). These migratory monocytes differentiate into macrophages of CNS (or perivascular macrophages) which may be responsible for productive HIV infection [Anderson *et al.,* 2002; Kaul *et al.,* 2001].

Microglial cells have been shown to be immuno-positive for HIV. *In vitro* studies have demonstrated that HIV replication takes place in primary microglial cells from adults [Albright *et al.*, 2008], infants [Ioannidis *et al.*, 1995] and fetal brain [McCarthy *et. al.*, 1998]. In CNS, microglial cells function in a manner similar to macrophages and express major receptors like CD4, CCR5 as well as other chemokine receptors *viz.*, CCR3, CCR2b, CCR8, CXCR6, and CX3CR1 [Martin-Garcia *et al.*, 2002].

Neurons

Neurons are non-permissive for HIV infections. Most of the studies have indicated absence of HIV infection *in vivo*; however a few studies have reported presence of HIV DNA and proteins in neurons [Bagasara *et al.*, 1996]. Absence of any receptors and co-receptors for HIV infection on neurons is considered as one of the strongest arguments for neurons' non-permissive nature. *In vitro* studies have reported a restricted infection to primary neurons [Ensoli *et al.*, 1995], and neuronal cell lines restricted to specific HIV strain, which failed to infect *in vivo* [Obregon *et al.*, 1999]. Other reasons for difficulty to demonstrate HIV infection in neurons is due to more complicated implication of inherent qualities of neurons. Because, in response to any adverse stimulus neurons tend to die either by apoptosis or by necrosis, which could be a reason for any/all failures to demonstrate HIV infectivity to neurons [Gonzalez-Scarano and Martin-Garcia, 2005].

Oligodendrocytes

Oligodendocytes are also CD4 negative cells, which raises a question, how these cells can get infected with HIV? This is one of the reasons in support of non-infectivity of oligodendrocytes. Some studies have demonstrated the presence of HIV nucleic acid in oligodendrocytes by *in situ* PCR. A limited HIV infectivity has also been reported *in vitro* [Albright *et al.*, 1996].

NEUROINVASION

Penetration of HIV into CNS is known as neuroinvasion. Neuroinvasion to CNS is still controversial because i) absence of known HIV receptors and co-receptors in most of the brain cells ii) CNS is protected by a unique protective layer known as blood-brain barrier (BBB) which acts as sentry for brain by regulating entry of any foreign entity to CNS, iii) any adverse response to neuron leads to neuronal cell death either by necrosis or by apoptosis.

Blood-brain barrier (BBB) is a continuous cellular layer of tightly linked microvascular endothelial cells of brain. BBB separates CNS from periphery (Figure 3). The permeability of BBB is selective in nature to regulate trafficking of cells and other substances which cross BBB. To enter into brain, any foreign material has to cross BBB, for which mechanism is still not clear. Numerous studies have used different animal models and *in vitro* experiments to understand the mechanisms of HIV introduction into the CNS through BBB [Gonzalez-Scarano and Martin-Garcia, 2005].

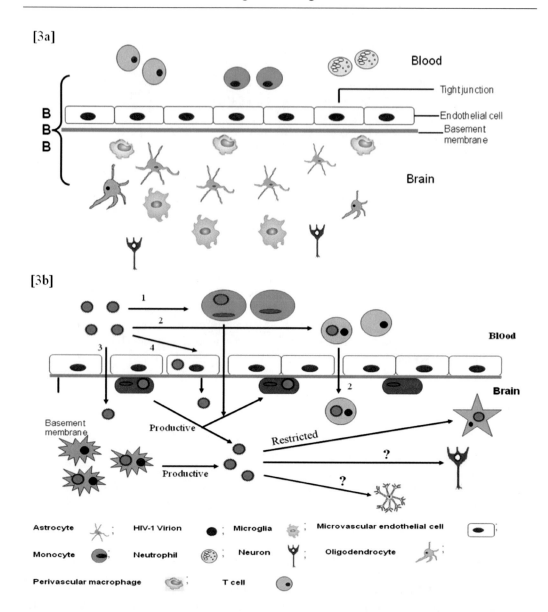

Figure 3. HIV-1 Neuroinvasion: This figure is a graphical representation of various cellular components in blood brain barrier (BBB). [3a] Represent a normal BBB barrier with its cellular components, without any breach. A BBB contains cells like astrocytes, microglia, microvascular endothelial cells, perivascular macrophages, oligodendrocytes, *etc.* [3b] Represent a BBB with its cellular components in case of HIV infection, where BBB is not intact. This figure is a graphical representation of "Trojan Horse Hypothesis" for HIV infection to CNS, where 1) HIV migrates to brain through infected monocytes and they differentiate into perivascular macrophages, 2) HIV infected T-cell can also be a source for HIV and provide a passage for HIV infection to brain, 3) Direct entry of HIV into brain through breach in BBB, 4) Entry of HIV into brain via microvascular endothelial cells by transcytosis and ?) Infection to neurons and oligodendrocytes is still questionable.

Although various models have been proposed for HIV infections, the most common one is "Trojan Horse Hypothesis" [Hasse *et al.*, 1986; Peulo *et al.*, 1985]. It is one of the most popular models to explain the HIV infection into CNS. As per this model HIV and other lentiviruses enter CNS as a passenger along with cells trafficked into the brain. Many HIV infected CD4$^+$ cells e.g. T-cells and monocytes which are infected with HIV remain in high numbers in blood circulation. These HIV infected cells enter into brain through BBB, where they help HIV to replicate and these new virion infect other cells of CNS viz., perivascular macrophages and microglial cells. The evidence to support this model came from in situ hybridisation, and immunohistochemical analysis (Takahashi et al., 1996). Though BBB abnormalities due to HIV infection have also been observed, however, the mechanisms of endothelial cells infection and the expression of conventional HIV receptors in these cells is still debatable. HIV infectivity of different CNS cell types have been shown in Figure 3b.

Apart from "Trojan Horse Hypothesis" model, some other models for HIV transmission into brain have been proposed with some scientific evidences, and they are i) direct entrance of virus into brain due to breach in BBB ii) entrance of HIV by transcytosis.

NEUROAIDS: CAUSATIVE FACTORS

NeuroAIDS is a debilitating disease, and it has so many causative factors. Contribution of these factors could be either individual, or a combination of them. However, a possibility of their synergistic effect cannot be denied. In general these causative factors can be divided into two main categories, i) intrinsic factors which includes host related factors, ii) extrinsic factors which includes virus and its components as well as anti-retroviral drugs.

1) Anti-retroviral Drugs

The medication used for HIV treatment, have also been implicated for their role in NeuroAIDS because of their side effects. Combination anti-retroviral therapy is so common nowadays that there are 25 approved anti-retroviral drugs in market with multitude of options [McCombe et al., 2009]. These can be used with simplified once-daily regimen. However, different drug regimens have varying degree of CNS penetrance. Therefore, CNS may act as reservoir for HIV. These ART may not provide complete protection against HIV infections to CNS. Several studies failed to show any correlation between progression of NeuroAIDS and different types of anti-retroviral drugs. Efavirenz is a NNRTI, commonly given to HIV patients. This drug is known to affect CNS when administered chronically. 50% patients treated with Efavirenz have been documented with its side effects related to CNS [Rihs et al., 2006; McCombe et al., 2009]. Other anti-retrovirals like didanosine and indinavir have also been reported to cause distal sensory neuropathy which causes neuropathic pain. Fortunately both of these drugs are not anymore in common use for HIV treatment, nowadays. Stauvudine (d4T) is part of WHO recommended ART regimen and widely used globally. Stauvudine causes neurological complications. A consistent increase in incidences of diabetes and dilipidemia among HIV patients has been also observed. The question remains, do these two clinical conditions also a contributing factor for NeuroAIDS? Chronic administration of different class of anti-retroviral drugs may have drug-to-drug interaction which could also initiate neuropsychiatric disorder.

2) High Risk Behavior

High risk behavior is very common among HIV patients. Is it not true that initial cases of HIV were reported among people with high risk behaviour like gay? An indulgence with substance of abuse, alcohol, sedatives, multiple sex partners, homosexuality, unprotected sex and prostitution make these HIV seropositives vulnerable to NeuroAIDS [Chin-Hong *et al.*, 2005].

3) Co-Infection and Co-Morbidities

A systemic immunosuppression is common among HIV seropositives even though they are under ART or HAART. A systemic immunosuppression increases the risk for opportunistic infections which may have serious consequences on nervous system. Some of the common clinical conditions are CNS infection, toxoplasmosis, encephalitis, cryptococcal meningitis, tuberculous meningitis, *etc.* These may cause seizures, physical and cognitive disabilities, psychosis and even mood disorders.

High risk behaviours among these patients make them vulnerable for increased risk of various co-infections, which may remain clinically undiagnosed for longer durations. These co-infections may have their own neuropsychiatric consequences. Commonly observed infections are syphilis and Hepatitis C. Hepatitis C is more common co-infection among HIV seropositives, which makes these patients highly vulnerable to HIV-associated dementia (HAD) [Brew, 2004]. Not only Hepatitis C itself, but therapies to treat Hepatitis C, like pegylated interferon-α, also have its own neuropsychiatric effect [Weiss *et al.*, 2006; Douiahy *et al.*, 2008].

We should not be surprised, if we find that increased incidences of NeuroAIDS in India are attributable towards high rates of incidences and wide spectrum of opportunistic infections. Incidences of opportunistic infections could be high due to tropical weather and bad sanitary conditions.

4) Virus Related Issues

Impact of HIV infection on host immune-system is detrimental. The long term effect of HIV infection is caused by latently infected cells. The latency of HIV in CNS is assumed to be one of the main contributing factors for neuropsychiatric disorder, as neurons are non-permissive for HIV infection. Some of the protective cells of CNS like microglia and monocyte derived macrophages (MDM) do get infected or remain latently infected. They may contribute towards neuropsychiatric disorder.

HAART has helped to control HIV infection; still it failed to cure it. It has been observed that viral replication goes below the detection limit; still a very low level of viral replication continues resulting in accumulation of different HIV virotoxins like gp120, tat, vpr, vpu, *etc.* in host circulation at chronic levels. These virotoxins may find entry into CNS by breaching BBB and can induce neurotoxicity. On one hand, some of these viral proteins reach to brain through blood circulation, while on another hand, latently infected or productively infected brain cells can also add different HIV virotoxins to brain. Out of these gp120, gp41, nef, tat and vpr are known to exert neurotoxic effects and apoptosis of neuronal cells [Ghafouri *et al.*, 2006].

5) Host Related Issues

HIV infected microglia/MDM produce neurotoxic soluble factors such as cytokines and chemokines, which induces neuronal injury, dendritic damage, and apoptosis. Indirectly these biomolecules induce inflammation via immune stimulation of uninfected astrocytes and microglia [Giulian *et al.*, 1990; Yeh *et al.*, 2000]. These factors include the TNF, IL-1β, and IL-6.

NEUROAIDS: INDIAN SCENARIO AND FUTURE PERSPECTIVE

There is a consistent rise in neurological and psychiatric complications among HIV seropositives, who have an access to better anti-retroviral treatments like HAART. Presently, in India, HIV seropositives have better access for anti-retroviral drugs, which helps to decrease morbidity and mortality among them. In near future, we expect to see rise in NeuroAIDS incidences in India. This warrants us to have a dialogue to control this health menace. At present, we need to develop programs like public awareness about NeuroAIDS. Healthcare personnels should be well trained to diagnose and treat NeuroAIDS at an early stage. This preparation will help us to reduce healthcare costs to treat these patients and will increase their productivity.

We even need to focus towards understanding biology of NeuroAIDS, which may help to develop better therapeutic strategies. An effort should be aimed to develop a neuroepidemiological database and prevalence of opportunistic infections. There is a high probability that flora of opportunistic infections in India may be different than opportunistic infections in other countries like USA, Canada, UK, *etc.* Different type of opportunistic infections may have a different impact on neuropsychiatric health complications. Study on development of neuropathological profiles/descriptions for primary and secondary HIV complications among different high risk groups like homosexuals, needle users, *etc.* are also essential. Even though HAART is very helpful to control HIV replication, still bioavailability of different drugs of HAART regimen to brain remains doubtful and in that case, it is essential to study drug resistance in HIV with reference to viral infections of CNS. Latency is another important concern for HIV infections to CNS, where we need to study activation and reactivation of latently infected cells of CNS. These are few important areas of investigation and concerns for NeuroAIDS to be addressed in near future in Indian context.

ACKNOWLEDGMENTS

Authors are thankful to Prof. A. K. Srivastava, Director, AIB, Amity University Uttar Pradesh, India for providing necessary resources and facilities for completion of this manuscript. Authors (ASV and AS) are thankful to one of their student Ms. Priyadarshini Mallick for her untiring efforts to perform literature searches. Authors are also thankful to Mr. Dinesh Kumar for his help for graphic designing and secretarial assistance.

REFERENCES

[1]   Albright, A.V., Strizki, J., Harouse, J.M., Lavi, E., O'connor, M. & Gonzalez-Scarano, F. (1996). HIV-1 infection of cultured human adult oligo endrocytes. *Virology,* 217: 211-219.

[2]   Albright, A.V., Shieh, J.T., O'Connor, M.J. & González-Scarano, F. (2000). Characterization of cultured microglia that can be infected by HIV-1. *J Neurovirol,* 6: S53-60.

[3]   Anderson, E., Zink, W., Xiong, H. & Gendelman, H.E. (2002). HIV-1-associated dementia: a metabolic encephalopathy perpetrated by virus-infected and immune-competent mononuclear phagocytes. *J Acquir Immune Defic Syndr,* 31: 43-54.

[4]   Bagasra, O., Lavi, E., Bobroski, L., Khalili, K., Pestaner, J.P., Tawadros, R.& Pomerantz, R.J. (1996). Cellular reservoirs of HIV-1 in the central nervous system of infected individuals: identification by the combination of in situ polymerase chain reaction and immunohistochemistry. *AIDS,* 10: 573-585.

[5]   Brack-Werner, R. (1999). Astrocytes: HIV cellular reservoirs and important participants in neuropathogenesis. *AIDS,* 13: 1-22.

[6]   Brew, B.J. (2004). Evidence for a change in AIDS dementia complex in the era of highly active antiretroviral therapy and the possibility of new forms of AIDS dementia complex. *AIDS,* 18: 75-78.

[7]   Chin-Hong, P.V., Deeks, S.G., Liegler, T., Hagos, E., Krone, M.R., Grant, R.M. & Martin, J.N. (2005). High-risk sexual behavior in adults with genotypically proven antiretroviral-resistant HIV infection. *J Acquir Immune Defic Syndr,* 40 : 463-471.

[8]   Dandona, L., Lakshmi, V., Kumar, G.A. & Dandona, R. (2006). Is the HIV burden in India being overestimated? *BMC Public Health,* 6: 308. {doi:10.1186/1471-2458-6-308}.

[9]   Dandona, L., Lakshmi, V., Sudha, T., Kumar, G.A. & Dandona, R. (2006). A population-based study of human immunodeficiency virus in south India reveals major differences from sentinel surveillance-based estimates. *BMC Med,* 4: 31. {doi: 10.1186/1741-7015-4-31}.

[10]   Douaihy, A., Hilsabeck, R.C., Azzam, P., Jain, A. & Daley, D.C. (2008). Neuropsychiatric aspects of coinfection with HIV and hepatitis C virus. *AIDS Read,* 18: 425-432.

[11]   Ensoli, F., Cafaro, A., Fiorelli, V., Vannelli, B., Ensoli, B. & Thiele, C.J. (1995). HIV-1 infection of primary human neuroblasts. *Virology,* 210: 221-225.

[12]   González-Scarano, F. & Martín-García, J. (2005) The neuropathogenesis of AIDS. *Nat Rev Immunol,* 5: 69-81.

[13]   Ghafouri, M., Amini, S., Khalili, K. & Sawaya, B.E. (2006). HIV-1 associated dementia: symptoms and causes. *Retrovirology,* 3: 28. {doi:10.1186/1742-4690-3-28}.

[14]   Giulian, D., Vaca, K. & Noonan, C.A. (1990). Secretion of neurotoxins by mononuclear phagocytes infected with HIV-1. *Science,* 250: 1593-1596.

[15]   Gorry, P.R., Ong, C., Thorpe, J., Bannwarth, S., Thompson, K.A., Gatignol, A., Vesselingh, S.L. & Purcell, D.F. (2003). Astrocyte infection by HIV-1: mechanisms of restricted virus replication, and role in the pathogenesis of HIV-1-associated dementia. *Curr HIV Res,* 4: 463-473.

[16] Gottlieb, M.S., Schroff, R., Schanker, H.M., Weisman, J.D., Fan, P.T., Wolf, R.A. & Saxon, A. (1981). Pneumocystis carinii pneumonia and mucosal candidiasis in previously healthy homosexual men: evidence of a new acquired cellular immunodeficiency. *N Engl J Med*, 305: 1425-1431.

[17] Haase, A.T. (1986). Pathogenesis of lentivirus infections. *Nature*, 322: 130-136.

[18] Hult, B., Chana, G., Masliah, E. & Everall, I. (2008). Neurobiology of HIV. *Int Rev Psychiatry*, 20: 3-13.

[19] Ioannidis, J.P., Reichlin, S. & Skolnik, P.R. (1995). Long-term productive human immuno-deficiency virus-1 infection in human infant microglia. *Am J Pathol*, 147: 1200-1206.

[20] Kaul, M., Garden, G.A. & Lipton, S.A. (2001). Pathways to neuronal injury and apoptosis in HIV-associated dementia. *Nature*, 410: 988-994.

[21] Martín-García, J., Kolson, D.L.& González-Scarano, F. (2002). Chemokine receptors in the brain: their role in HIV infection and pathogenesis. *AIDS*, 16: 1709-1730.

[22] McArthur, J.C., Brew, B.J., Nath, A. (2005). Neurological complications of HIV infection. *Lancet Neurol*, 4: 543-555.

[23] McCarthy, M., He, J. & Wood; C. (1998). HIV-1 strain-associated variability in infection of primary neuroglia. *J Neurovirol*, 4: 80-89.

[24] McCombe, J.A., Noorbakhsh, F., Buchholz, C., Trew, M. & Power, C. (2009). NeuroAIDS: a watershed for mental health and nervous system disorders. *J Psychiatry Neurosci*, 34: 83-85.

[25] Obregón, E., Punzón, C., Fernández-Cruz, E., Fresno, M. & Muñoz-Fernández, M.A. (1999). HIV-1 infection induces differentiation of immature neural cells through autocrine tumor necrosis factor and nitric oxide production. *Virology*, 261: 193-204.

[26] Peluso, R., Haase, A., Stowring, L., Edwards, M. & Ventura, P. (1985). A Trojan Horse mechanism for the spread of visna virus in monocytes. *Virology*, 147: 231-236.

[27] Rihs, T.A., Begley, K., Smith, D.E., Sarangapany, J., Callaghan, A., Kelly, M., Post, J.J. & Gold, J. (2006). Efavirenz and chronic neuropsychiatric symptoms: a cross-sectional case control study. *HIV Med*, 7: 544-548.

[28] Takahashi, K., Wesselingh, S.L., Griffin, D.E., McArthur, J.C., Johnson, R.T. & Glass, J.D. (1996). Localization of HIV-1 in human brain using polymerase chain reaction/in situ hybridization and immunocytochemistry. *Ann Neurol*, 39: 705-711.

[29] Verma, A.S., Bhatt, S.M., Singh, A. & Dwivedi, P.D. (2008). HIV: An Introduction. In: Chauhan AK, Varma A editor. *Text Book on Molecular Biotechnology*. Delhi: I.K. International; 853-878.

[30] Weiss, J.J. & Gorman, J.M. (2006). Psychiatric behavioral aspects of co-management of hepatitis C virus and HIV. *Curr HIV/AIDS Rep*, 3: 176-181.

In: Recent Trends in Biotechnology and Microbiology   ISBN: 978-1-60876-666-6
Editors: Rajarshi Kumar Gaur et al., pp. 169-179   2010 Nova Science Publishers, Inc.

Chapter 14

IN VITRO MORPHOGENESIS (RHIZOGENESIS) IN KALIHARI (*GLORIOSA SUPERBA* L.)- AN ALTERNATIVE SOURCE OF COLCHICINE PRODUCTION

M.S. Rathore[1], Mangal S. Rathore[2], D. Panwar[1] and N.S. Shekhawat[1]

Biotechnology Unit, Jai Narain Vyas University,
Jodhpur (Rajasthan-342033) India[1]
Biotechnology Laboratory, Department of Science, FASC,
MITS Deemed University, Lakshmangarh, Sikar (Rajasthan- 332311)[2]

ABSTRACT

Gloriosa superba L., a member of the Liliaceae family is locally called as "Kalihari". Plant produces alkaloids, mainly colchicine and colchicoside. The derivatives of colchicines have shown promising anti-cancer and anti-inflammatory properties. The normal regeneration and propagation of Gloriosa superba is through corm are slow and insufficient for conservation. Poor seed germination, susceptibility towards many pests, collection of seeds by experts for pharmaceutical industries and reduction in forest area are important factors responsible for diminishing population. Seeds of G. superba germinated on MS medium + 0.1 mg/l BAP. The activated embryo exhibited callus induction within 23-25 days from inoculation on MS medium + 2,4-D (2.0 mg/l). MS medium supplemented with 1.0 mg/l 2,4-D + 0.5 mg/l Kinetin + 0.5% Maltose + additives was found to be the best medium for multiplication of cell cultures. On MS + 1.0 mg/l BAP + 0.5 mg/l Kinetin callus turned embryoidogenic. On MS medium + 2.0 mg/l BAP + 0.5 mg/l Kinetin the callus showed rhizogenesis. The methodology developed and defined is reproducible and can be utilized for production of active constituent (colchicine) using plant biotechnology tools.

*Email- mahendersr@gmail.com

INTRODUCTION

Gloriosa superba L., a member of the Liliaceae family, locally known as "Kalihari", is an over exploited ornamental and important medicinal plant in India. Native of tropical Asia and Africa is found growing throughout the tropical parts of India and is the only species representing the genus. In nature the shoots dry up every year and the corm perennate and subsequently produce new shoots and flowers (Figure 1). Flowering occurs after 3-4 years when the size of corm has increased to a critical size specific to the species (Shivkumar *et al.*, 2004). The plant has been used to cure various ailments in the traditional Ayurvedic medicine since ancient times (Shivkumar and Krishanamurthy, 2004). *Gloriosa superba* produces alkaloids, mainly colchicine and colchicoside (Shivkumar, 2003). The main uses of colchicines include chromosome manipulation, treatment of familial Mediterranean fever, treatment of gout (Sivakumar, 2003) and skin and lever diseases (Shivkumar and Krishanamurthy, 2002). The derivatives of colchicines have shown promising anti-cancer and anti-inflammatory properties. The normal regeneration and propagation of *Gloriosa superba* is through corm are slow and insufficient for conservation (Shivkumar and Krishanamurthy, 2002), (Ghosh and Jha, 2004). Poor seed germination, susceptibility towards many pests, collection of seeds by experts for pharmaceutical industries and reduction in forest area are important factors responsible for diminishing population and therefore has been declared endangered (Ghosh *et al.*, 2002). Pharmacies and drug manufacturers often fill up to 75% of there raw material demand of this species from the wild. Of this 75%, about 60% comes from destructive collection, which includes either the entire plant or underground parts like corms and roots or reproductive parts like fruits and seeds (Shivkumar and Krishanamurthy, 2002). Thus, mass multiplication through tissue culture is needed not only to conserve this taxon but also to meet the demands for this medicinal plant as a source of colchicines (Finnie and Staden, 1989).   For commercial production of colchicine and its derivatives, natural production from *in vitro* methods is of great interest. In the past two decades focus has been on plant cell biotechnology as a possible alternative production method, using cultured cells rather than plants (Shivkumar and Krishanamurthy, 2004; Ghosh and Jha, 2004).

Figure 1. Attractive flower of *G. superba*.

Gloriosa superba callus culture may be an alternative source of phyto-colchicine production (Shivkumar and Krishanamurthy, 2002). *In vitro* production of secondary metabolites has many advantages like (i) year-round availability of the plant material for the production of functional phyto-molecules (ii) better avenues for processing and isolation, and (iii) the possibility of accentuation of chemical reaction leading to useful secondary metabolites under *in vitro* conditions. The application of *in vitro* techniques offers the possibility of producing large number of uniform plantlets. Cell and root cultures have been successfully exploited for biotechnological production of several alkaloids. Although undifferentiated cell cultures do not often produce alkaloids because alkaloids synthesis is linked to root differentiation (Ghosh and Jha, 2004; Finnie and Staden, 1989).

MATERIALS AND METHODS

Explant Selection, Surface Sterilization and Inoculation

Plants of *G. superba* along with mature tubers were collected from Aravalli region of Udaipur division. These source plants were then maintained and managed in a green house at plant biotechnology Laboratory, Department of Botany, J.N.V. University, Jodhpur. Various types of explants were utilized to establish the cell cultures like-activated embryo, leaf base, flower bud and apical bud. The seeds of *G. superba* were treated with 0.1% Bavistin for 12-15 minutes followed by surface sterilization with 0.1% $HgCl_2$ solution for 3-4 minutes. These were given 6-8 washing with sterile water. Surface sterilized seeds were inoculated on MS (Murashige and Skoog, 1962) semisolid medium supplemented with 0.1mg/l BAP + additives (50.0 mg/l ascorbic acid, 25.0 mg/l each of citric acid and adenine sulphate) for the activation of embryo.

Young flowerer buds were harvested from plants in the months of September-October. After surface sterilization with $HgCl_2$ solution and washing with sterilized water, the buds were cut into small pieces and were inoculated on MS medium supplemented with 2,4-D (2.0mg/l). Tubers sprouted during rainy season's i.e in the months of July-August were also used as explants for the present study. The tuber were firstly washed with 5% (v/v) solution of liquid detergent and then with 0.1% of Bavistin for 20-25 minutes, followed by surface sterilization with 0.1% $HgCl_2$ solution for 3-4 minutes. These were given 6-8 washing with sterile water. Subsequently 1.0cm long portion of the tuber containing the apical bud were dissected aseptically to raise *in vitro* cultures on 2,4-D (2.0mg/l). The activated embryos were inoculated on MS medium + 2,4-D (1.0-10.0 mg/l) or 2,4,5-T or Dicamba (0.25-3.0 mg/l). The cultures were incubated in growth chamber at $26\pm2°C$, 60% RH under diffused light (15-20μmolm^{-2}s^{-1} Spectral Flux Photon, SFP) and 12 hd^{-1} photoperiod.

Multiplication of Cell Cultures

The callus initiated from explants was separated and subcultured on MS medium with various concentrations of 2,4-D (1.0-5.0 mg/l) alone or in combination with cytokinins (BAP/Kinetin) at concentrations 0.1-1.0 mg/l along with 0.5% Maltose and additives. About 100.0 mg of callus was subcultured on to fresh medium after every 2-3 weeks.

Organogenesis in Cell Cultures

For regeneration the organogenesis responsive callus was further transferred on plant growth regulator free-MS medium + 0.5% Maltose. The callus was also inoculated on MS + varied concentrations of BAP (1.0-5.0 mg/l) or Kinetin (1.0-5.0 mg/l) or combination of both BAP (1.0-5.0 mg/l) + Kinetin (0.5 mg/l) + 0.5% Maltose + additives along with activated charcoal.

RESULTS AND DISCUSSION

Establishment of Tissue Cultures

Seeds of *G. superba* germinated on MS medium + 0.1 mg/l BAP. The seeds with intact seed coat did not germinate *in vitro*. More than 40% of the cut seeds cultured on the above medium germinated within 15 days (Shivkumar and Krishanamurthy, 2004). A 3-4 week old seedling growing on this medium has one cotyledon and a root. Among all types of explants tested for establishment of culture, activated embryo was found the most suitable (Table 1and Figure 2). The activated embryo was responded and produced compact, mucilaginous and moderately growing callus (Shivkumar *et al.*, 2004; Shivkumar, 2003; Shivkumar *et al.*, 2003). The activated embryo exhibited callus induction within 23-25 days from inoculation on MS medium + 2, 4-D (2.0 mg/l). Other explants like leaf base and apical bud was less responsive and produced dark brown, non-regenerative callus. The flower bud produced soft, friable and slow growing callus. Activated embryo produced callus on MS medium supplemented with 2,4-D (1.0-10.0 mg/l), 2,4,5-T and Dicamba (0.25-3.0 mg/l) + additives (Figure 3). Activated embryo produced compact, slight mucilaginous and regenerative callus on MS + 2.0 mg/l 2, 4-D. Maximum response was observed on this medium (Table 2).

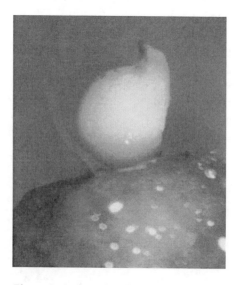

Figure 2. Activated embryo of G. *superba*.

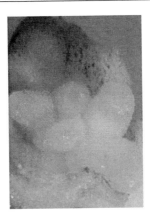

Figure 3. Callus initiated from activated embryo.

Table 1. Responses of explants of *G. superba* on MS medium + 2.0 mg/l of 2, 4-D.

S.No.	Explant types	Response	Remarks
1.	Activated embryo	++	Produced compact, mucilaginous callus
2.	Leaf base from *in vitro* grown seed	+	Produced soft, friable callus
3.	Flower bud	++	Produced dark brown callus
4.	Apical bud	+	Produced dark brown callus

+ = slight callusing, ++ = moderate callusing, +++ = fast growth of callus

Table 2. Effect of different concentrations of 2,4-D on callus induction from the activated embryo on MS + additives.

S.No.	Conc. of 2,4-D (mg/l)	Callus intensity	Remarks
1.	0.0	-	No response
2.	1.0	+	Slow callusing
3.	2.0	++	Creamy, moderately growing, compact and slightly mucilaginous callus
4.	3.0	++	Fragile, creamy white callus
5.	4.0	++	Slow growing, watery callus
6.	5.0	+	Slow callusing, browning of callus
7.	6.0	+	Browning of callus
8.	7.0	-	No response
9.	8.0	-	No response
10.	9.0	-	No response
11.	10.0	-	Browning of embryo

+ = slight callusing, ++ = moderate callusing, +++ = fast growth of callus

Callus obtained from activated embryo was initially covered with mucilage which disappeared after 2-3 subcultures. The callus initiation was maximum and fast on 2.0 mg/l of 2,4-D. On higher concentration of 2,4-D the callus initiated turned brown and become non-regenerative. On medium supplemented with 2,4,5-T or Dicamba, the callus initiation was late and also the amount of callus obtained was too small (Table 3). On medium containing 3.0 mg/l of 2,4,5-T, browning of explant observed while similar concentration of Dicamba was found to be toxic for activated embryo.

Table 3. Effect of different concentrations and types of auxins on callus induction from the activated embryo on MS + additives.

PGR	Concentration (mg/l)	Response	Remarks
2,4,5-T		+	Slow growing, glassy callus
	0.25	-	No response
	0.50	-	No response
	1.0	-	No response
	2.0	-	No response
	3.0	-	Browning of inoculum
Dicamba			
	0.25	+	Very little , fragile callus
	0.50	-	No response
	1.0	-	No response
	2.0	-	Embryo turned brown
	3.0	-	Embryo turned brown and ultimately died.

+ = slight callusing, ++ = moderate callusing, +++ = fast growth of callus

Multiplication of Cell Cultures

The callus initiated on MS + 2.0 mg/l 2,4-D, started to become brown on fresh medium after 4-5 passage. Therefore the callus was transferred on MS medium supplemented with lower (0.0-5.0 mg/l) concentration of 2,4-D. Callus initiated from activated embryo was transferred to MS medium supplemented with lower (1.0 mg/l) concentration of 2,4-D + 0.5 % Maltose + additives. On this combination the callus produced was compact, moderate growing; regenerative and non-mucilaginous (Table 4).

Incorporation of Cytokinins (BAP and Kin) promoted the rate of callus production. MS medium supplemented with 1.0 mg/l 2,4-D + 0.5 mg/l Kinetin + 0.5% Maltose + additives was found to be the best medium for multiplication of cell cultures (Table 5). On this medium the callus produced was fast growing, regenerative, organized and produced globular structures (Shivkumar et al., 2003; Shivkumar and Krishanamurthy, 2002). On higher concentration (more than 0.5) of BAP and Kin, the callus produced was green in color, while on lower (less than 0.5 mg/l) concentration the rate of callus production was slow. Kinetin alone was found to be most suitable for multiplication of culture as compared to BAP alone and in combination with BAP.

Table 4. Effect of concentration of 2,4-D on callus multiplication of *G. superba* on MS medium + 0.5% Maltose + additives.

S. No.	Conc. of 2,4-D (mg/l)	Callus multiplication rate (per 100.0 mg of inoculation)	Remarks
1.	0.0	One fold	Very slow growth
2.	1.0	Three fold	Compact, moderate growing, regenerative callus and non-mucilaginous
3.	2.0	Two fold	Creamy slow growing callus
4.	3.0	Two fold	Soft white callus
5.	4.0	Two fold	Watery, unorganized callus
6.	5.0	One fold	Slow growing callus

+ = slight callusing, ++ = moderate callusing, +++ = fast growth of callus

Table 5. Effect of types and concentrations of cytokinins on callus multiplication of *G. superba* on MS medium + 1.0 mg/l 2,4-D + 0.5 % Maltose + additives.

PGR	Concentration (mg/l)	Callus intensity	Remarks
Control	0	-	No response
BAP			
	0.1	+	Soft watery callus
	0.2	+	Slow growth
	0.3	++	Slow growth
	0.5	+	Greenish callus
	0.8	+	Callus started to become green
	1.0	+	Greening of callus
Kinetin			
	0.1	+	Slow growth
	0.2	+	Slow growth
	0.3	+	Moderate growth and regenerative callus
	0.5	+++	4-5 fold, fast growth, regenerative, organized/globular structures seen
	0.8	++	Light green callus
	1.0	+	Greening of callus
BAP + Kinetin	0.25 each	++	Light green callus
BAP + Kinetin	0.50 each	+++	3-4 fold, callus glossy and green

+ = slight callusing, ++ = moderate callusing, +++ = fast growth of callus

Organogenesis in Cell Cultures

On hormone free-MS medium callus showed slight growth but did not respond in terms of organogenesis. On MS medium + BAP the response was poor. On MS + 1.0 mg/l BAP slight embryoidogenesis was observed with rhizogenesis (Shivkumar *et al.*, 2003). On higher concentration of BAP, callus started to become green and glossy. On MS medium with lower concentration of Kinetin pale yellow callus was formed. On higher concentration of Kinetin very slow response was observed. On MS + 1.0 mg/l BAP + 0.5 mg/l Kin callus turned embryoidogenic (Figure 4). On this media composition the callus produced was organized and globular. On MS + 2.0 mg/l BAP + 0.5 mg/l Kin + 0.1% activated charcoal, the callus showed rhizogenesis (Figure 5). Increase in BAP concentration inhibited rhizogenesis (Table 6). It is common observation that auxins promote root formation (Wilson and Staden, 1990; Schiefelbein and Benfey, 1991) but, here cytokinins were found to me more responsive. It can be due to interaction of exogenously supplied plant growth regulators (PGRs) with internal PGRs (Kuppusamy *et al.*, 2009; Fukaki and Tasaka, 2009).

Figure 4. Embryoids produced from callus.

Figure 5. Rhizogenesis in *G. superba* cell cultures.

Table 6. Effect of types and concentrations of cytokinins on organogenesis responses of G. superba on MS medium + 0.5 % Maltose + additives.

PGR	Concentration (mg/l)	Remarks
Control	0	No response in terms of organogenesis
BAP	1.0	Slight embryoidogenesis with rhizogenesis
	2.0	Greening of callus
	3.0	Callus turned green
	4.0	Green callus
	5.0	Callus turned green
Kinetin	1.0	Pale yellow callus
	2.0	Embryoids formed
	3.0	Embryoids formed but rhizogenesis observed
	4.0	No regeneration
	5.0	No regeneration
Kinetin 0.5 + BAP	1.0	Embryoidogenesis
	2.0	Rhizogenesis
	3.0	Less rhizogenesis
	4.0	No rhizogenesis / no regeneration
	5.0	No rhizogenesis/ no regeneration

CONCLUSIONS

Protocol described here for *G. superba* is successful and reproducible, can be used for germplasm conservation. From the present study it can be concluded that, out of all types of explants tested for establishment of culture, activated embryo was found the most suitable. The activated embryo exhibited callus induction. Incorporation of activated charcoal in the culture medium is suggested for amplification of cultures, as it adsorbs the phenolic compounds secreated by damaged cells or dead cells, which are deleterious for the plant growth (Thomas, 2008; Thomas and Shankar, 2009). Callus obtained from activated embryo was initially covered with mucilage, which disappeared after 2-3 subcultures. This mucilaginous covering can be removed manually to fasten the process of culture amplification. Rhizogenous cultures produced can be useful for extraction of active compounds like colchicines, without the need of *Agrobacterium rhizogenes* for the induction of hairy root cultures.

ACKNOWLEDGMENTS

Research programs in the Department of Botany are sponsored by the University Grants Commission (UGC) India. Micropropagation and Hardening facilities are established by the financial support from the Department of Biotechnology (DBT), Govt. of India. The Department of Science and Technology (DST), Govt. of India provided funds for infrastructure development in the Department of Botany under FIST programme.

REFERENCES

[1]   Finnie, J.F. & Staden, J.V. (1989). In vitro propagation of Sandersonia and Gloriosa. *Plant Cell Tissue and Organ Culture*, 19: 151-158.

[2]   Fukaki, H. & Tasaka. (2009). Hormone interactions during lateral root formation. *Plant Molecular Biology*, 69: 437-449.

[3]   Ghosh, B., Mukherjee, S., Jha, T.B. & Jha, S. (2002). Enhanced colchicine production in root cultures of Gloriosa superba by direct and indirect precursors of the biosynthetic pathway. *Biotechnology Letters*, 24: 231-234.

[4]   Ghosh, S. & Jha, S. (2004). Colchicine accumulation in in vitro cultures of Gloriosa superba L. In: *Biotechnology for a better future*. SAC Publications, Manglore.

[5]   Kuppusamy, K.T., Walcher, C.L. & Nemhauser. (2009). Cross-regulatory mechanisms in hormone signaling. *Plant Molecular Biology*, 69: 375-381.

[6]   Murashige, T. & Skoog, F. (1962). A revised medium for rapid growth and bioassays with tobacco tissue cultures. *Physiologia Plantarum*, 15: 473-497.

[7]   Schiefelbein, J.W. & Benfey, P.N. (1991). The development of plant roots: New Approaches to underground problems. *Plant Cell*, 3: 1147-1154.

[8]   Shivkumar, G. (2003). Conventional and In vitro Propagation of the Medicinal Plant Gloriosa superba, *AgriSupportOnline*, 1-3.

[9]   Shivkumar, G. & Krishanamurthy, K.V. (2002). Micropropagation of Gloriosa superba – an over exploited medicinal plant species from India. In: Nanda, S.K., Palni, L.M.S. and A. Kumar (eds.), *Role of plant tissue culture in biodiversity conservation and economic development*, Gyanodaya Prakashan Pub. Nainital, India. pp. 345-350.

[10]   Shivkumar, G. & Krishanamurthy, K.V. (2004). In vitro Organogenetic Responses of Gloriosa superba. *Russian Journal of Plant Physiology*, 51: 790-798.

[11]   Shivkumar, G., Krishanamurthy, K.V. & Rajendran, T.D. (2003). Embryoidogenesis and plant regeneration from leaf tissue of Gloriosa superba. *Planta Medica*, 69: 479-481.

[12]   Shivkumar, G., Krishanamurthy, K.V., Hao, J.A. & Paek, K.Y. (2004). Colchicine production in Gloriosa superba calluses by feeding precursors. *Chemistry of Natural Compounds*, 40: 499-502.

[13]   Thomas, T.D. (2008). The role of activated charcoal in plant tissue culture. *Biotechnology Advances*, 26: 618-631.

[14]   Thomas, T.D. & Shankar, S. (2009). Multiple shoot induction and callus regeneration in Sarcostemma brevistigma Wight & Arnott, a rare medicinal plant. *Plant Biotechnology Reports*, 3: 67-7

[15] Wilson, P.J. & Staden, V. J. (1990). Rhizocaline, Rooting Co-factors, and the Concept of Promoters and Inhibitors of Adventitious Rooting-*A Review. Annals of Botany*, 66: 479-490.

In: Recent Trends in Biotechnology and Microbiology
Editors: Rajarshi Kumar Gaur et al., pp. 181-188

ISBN: 978-1-60876-666-6
2010 Nova Science Publishers, Inc.

Chapter 15

GENETICALLY MODIFIED ORGANISMS AND THEIR APPLICATIONS IN MODERN ERA

Sakshi Issar and Harish Dhingra

Departmet of Science, Mody Institute of Technology & Science,
Laxmangarh, Distt. Sikar, (Rajasthan)

ABSTRACT

The manipulation of living beings has undergone a real revolution with the arrival of genetic engineering, which allows part of the gene pool to be isolated and manipulated as desired. Thus, genetic engineering has led to the appearance of "genetically modified organisms" or GMOs. A Genetically modified organism (GMO) or genetically engineered organism (GEO) is an organism whose genetic material has been altered using genetic engineering techniques. These techniques are generally known as recombinant DNA technology. With the recombinant DNA technology, DNA molecules from different sources are combined in vitro into one molecule to create a new gene. This DNA is then transferred into an organism and causes the expression of modified or novel traits. Contemporary genetic modification was developed in the 1970s and essentially transfers genetic material from one organism to another.

INTRODUCTION

With the development of recombinant DNA technology, the metabolic potentials of microorganisms are being explored and harnessed in a variety of new ways. GMMs mean bacteria, yeasts, filamentous fungi in which genetic material have been changed through modern biotechnology in a way that does not occur naturally by multiplication and natural recombination. Here modern biotechnological methods imply:

- In vitro nucleic acid techniques, including recombinant deoxyribonucleic acid (DNA) and direct injection of nucleic acid into cells or organelles, or

- Fusion of cells beyond the taxonomic family that overcome natural physiological, reproductive or recombination barriers and that are not techniques used in traditional breeding and selection.

Microorganisms have been developed which may be characterized as possessing substantially all of the following qualities or capabilities:

a) Capable of having foreign genetic information introduced there into and recovered there from along with its expression with production of useful gene products;
b) The microorganism being dependent for growth and survival upon defined conditions;
c) The microorganism being incapable of establishment or growth or colonization and/or survival under conditions or in ecological niches that are considered to be natural and/or undesirable for said microorganism;
d) The microorganism being capable of causing genetic information incorporated therein to undergo degradation under conditions or ecological niches that are considered to be natural and/or undesirable for said microorganism;
e) The microorganism being capable of permitting cloning vectors incorporated therein to be dependent for their replication, maintenance and/or function on said microorganism;
f) The microorganism being substantially incapable of transmitting cloning vectors or recombinant DNA molecules incorporated therein to other organisms under conditions or ecological niches that are considered to be natural and/or undesirable for said microorganism;
g) The microorganism being capable of being monitored by suitable means and/or techniques without substantial alteration of said microorganism;
h) The microorganism being susceptible of substantially minimal contamination with other organisms when recombinant DNA molecules are incorporated therein and being substantially incapable of contaminating other organisms when incorporated therein or consumed thereby when recombinant DNA molecules are incorporated in said microorganism.

The first transgenic microorganism to be produced was bacterial strain *E. coli* in 1973 expressing a *Salmonella* gene. The microorganisms are genetically engineered in several ways such as transformation, electroporation etc but the basic principle states that the selected genetic material is introduced into recipient microorganisms by recombinant DNA techniques. This genetic material may be integrated into the resident chromosomes of the cells or extrachromosomal structures like plasmids. The genetic information is then transferred through a process of replication, cell division and chromosomal segregation. GMOs are referred to as transgenic, because they contain the DNA of more than one organism (Snow, 1999 & 2003).

Today genetically modified microorganisms (GMMs) have found application in human health, agriculture, and bioremediation and in industries such as food, paper, textiles etc. In addition, genetic engineering offers sufficient supplies of desired products, cheaper product production and safe handling of otherwise dangerous agents. Genetically modified microorganisms (GMMs) are being developed for a very wide range of applications. Some are intended for use in contained situations only and are already used on a commercial scale, e.g. GM microorganisms that produce insulin for diabetics. Others, however, are intended for use in the open environment. Many of the GMMs have applications in agriculture (for pest or disease control), in polluted environments as bioremediation agents or in the control of insect-borne diseases of humans. Commercial development of these GM microorganisms is not as far advanced as that of the contained GM microorganisms or of GM plants (Bakan, *et al.* 2002). The potential non-target impacts of GM microorganisms will differ according to their biological activity and their intended environment. Potential environmental impacts of GM microorganisms will also depend on the extent of spread from the site of release (Conner, *et al.* 2003).

APPLICATION OF GMM IN MODERN ERA

Gmms in Agriculture

Biological Control of Frost Injury in Plants
Frost damage is a major agricultural problem affecting many annual crops such as deciduous fruit trees; subtropical plants etc. in addition to the losses caused by frost injury, hundreds of millions of dollars are spent every year to reduce plant frost injury mechanically. As a result, these methods are both costly and ineffective. Bacteria belonging to the genera *Pseudomonas, Xanthomonas*, initiate Frost damage and *Erwinia,* which are collectively, called ice-nucleating bacteria. The bacteria living on the surface of the plants possess a membrane protein that acts an ice nucleus for initiation of ice crystal formation. The ice crystals so formed disrupt plant cell membranes, thus causing cell damage. The biological route of controlling the nucleating bacteria is through seed or foliar applications of non-ice nucleating bacteria to outcompete ice-nucleating bacteria. Treating the ice-nucleating bacteria with chemical mutagens isolated the non ice-nucleating bacteria. Chemically induced mutants suffer from a disadvantage that they often harbour multiple mutations that may adversely impact their genetic stability and ecological fitness. Therefore, to avoid multiple mutations, ice-nucleating deficient mutants of *Pseudomonas syringae* were constructed by deleting the genes conferring ice nucleation. These genetically engineered mutants were able to compete successfully with ice-nucleating *P. syringae*. Further, field tests showed that plants treated with ice-nucleation deficient *P. syringae* suffered significantly less frost damage than the untreated control plants (Munkvold, *et al.* 1997).

Biological Control of Insect Pests
Bacillus thuringiensis (Bt) is a naturally occurring soil-dwelling sporulating bacterium, produces unique crystal like proteins that have larvicidal activities against different insect species and it poses no threat to mammals, birds or fish. These crystal like proteins are produced by Bt spores, when cleaved in the alkaline mid-gut of the insect bind to specific receptors on mid gut epithelial cells, causing cell lysis and death (Carpenter, 2001). Because of its unique features, Bt has been used as a safe alternative to chemical pesticides for several decades. However, natural Bt based products do possess some shortcomings, including instability in the natural environment, narrow host range, need for multiple application and

difficulty in reaching the crops internal region where various larvae feed (Malone, & Pham-Delegue, 2001).

In order to overcome these problems plant-associated bacteria are used as hosts for delivering the toxins. Bt toxin genes have been introduced successfully into several plant-associated bacteria, including *Clavibacter xyli* subsp. *cynodontis* and *Ancylobacter aquaticus* (Wang, 2003, Yamaguchi, & Blumwald, 2005). Genetically modified cynodontis containing the Bt toxin gene integrated into the chromosome showed moderate control of European corn borer. Also the modified *A. aquaticus* strain expressing Bt toxin genes, introduced by electroporation, exhibited significant toxicity towards mosquito larvae, thus demonstrating its potential in mosquito control (Glare & O'Callaghan, 2000).

There have been a number of viral-based biopesticides. One of the best known is Gypchek, a viral pesticide in the form of a specific nucleopolyhedrosis virus developed for the control of gypsy moth, *Lymantria dispar*. The nucleopolyhedrovirus (LdNPV) is a specific virus, which attacks only *Lymantria* spp. making it one of the most specific agents available. It occurs as a natural control agent in most gypsy moth populations, but is also formulated as a product in several countries (Crailsheim, 1990).

Biological Control of Plant Disease

Plant pathogens such as fungi, viruses, and bacteria damage crops and thereby reduce crop yield. Plant diseases are conventionally fought with chemicals, a strategy that is expensive, inconvenient, potentially environment unfriendly and sometimes ineffective. An alternative method is to develop biological control agents in which microorganisms are modified to deliver the desired chemicals (Munkvold, *et al.* 1999).

Agrobacterium tumefaciens causes crown gall disease in a wide range of broad-leaved plants by transferring part of its DNA (T-DNA), located on a large tumour-inducing (Ti) plasmid, into the plant cell. Upon integration of T-DNA into the plant hosts chromosome, the genes on T-DNA are expressed, resulting in overproduction of plant growth hormones. Overproduction of plant growth hormones causes cancerous growth. *Agrobacterium radiobacter* K84 produces a bacteriocin, agrocin 84, to which pathogenic *A. tumefaciens* are susceptible (Gianessi, 2005). Therefore, *A. radiobacter* K84 became the first commercial biological control agent against crown gall disease (Dale, *et al.* 2002, Romeis, 2006).

Other plant disease control methods using GMMs include:

1) Overproduction of oomycin A
2) Synthesis of phenzine-1-carboxylic acid in a heterologous host
3) Heterologous expression of a lytic enzyme gene.
4) Phlebiopsis gigantea for control of stem and root rot of pines and *Trichoderma* spp. for control of a range of diseases.
5) Two commercialised products are based on fungi, *Myrothecium verrucaria* and *Paelicomyces lilacinus*, and two on bacteria; *Burkholderia cepacia* and *Pasteuria penetrans* the products have broad-spectrum activity against nematodes in soil.

Soil Amendment

Genetic modification in microorganisms to improve soil fertility has also been developed. *Medicago sativa (alfalfa)*, grown in soils with a high nitrogen concentration, has been shown to undergo better root nodulation when exposed to a genetically modified *Sinorhizobium*

(Rhizobium) meliloti expressing the *Klebsiella pneumonia nifA* gene than plants in the same environment exposed to wild type *S.meliloti*. According to another study, the recombinant *S.meliloti* significantly increased plant biomass when compared to the wild type strain (James, 2004 & James, 2005).

Biofuel Production

Researchers have been able to develop a method of producing biofuels by genetically modifying the *E. coli* bacteria and making it into an efficient biofuel synthesizer. In *E. coli* strain of bacteria, key pathways have been modified to make them produce higher chain alcohols out of glucose. These alcohols are the typical trace by-products of fermentation. The new strategy developed has opened up a new method of producing biofuels with the use of bacteria and other microorganisms.

Currently, biofuels are being produced through the processing of agricultural products such as corn and sugarcane. Using current processes, the biofuel ethanol has largely been produced but has been known to be inefficient when mixed with gasoline. In order to be more efficient, this biofuel may need to be mixed with a gas in order to be used as a transportation fuel. Ethanol also tends to absorb water in its surroundings and may promote corrosion not only in car engines but also in the structures built to store them. Therefore, biofuels produced with higher chain alcohols such as isobutanol provide a better option and may have closer energy densities to that of gasoline. These next generation biofuels are also not as volatile and/or corrosive as ethanol nor do they absorb water as the current biofuel ethanol. Isobutanol and other C5 alcohols also have higher octane numbers and may also offer less knocking in engines. The study of using genetically modified *E. coli* bacteria to produce such higher chain alcohols have provided the viable means from which the next generation of biofuels can be produced commercially in the future.

Gmms as Diagnostic Tools

Acquired immunodeficiency syndrome (AIDS) immunological tests are used for diagnosing the disease and for testing donated blood samples. The first generation of AIDS tests, commercialized in 1985, was based on inactivated human immunodeficiency virus (HIV) grown in tissue culture. This production method is both expensive and, more importantly, hazardous because of the risk from handling the infectious agent. Further, this first generation AIDS test was subject to false-positive reactions because of the cellular debris from virus producing human cells. These problems were overcome by cloning the gene encoding the relevant antigenic coat protein of the virus into *E. coli* for large-scale production of the protein. Other diagnostic tests that have been developed using GMMs include one for diagnosing Alzheimer's disease. Noninvasive diagnosis of Alzheimer's disease was not possible until the development of an enzyme-linked immunosorbent assay kit in the mid 1990s.

Gmms in Textile Industry

Microbial enzymes have been used in the textile industry since the early 1900s. For commercialization, enzymes must be produced at high levels. Conventional methods to enhance production include optimizing medium composition, growth condition and the fermentation process. Random mutagenesis and screening commonly is used to achieve the required high yields. Genetic engineering offers a possibility in which high level enzyme production is achieved in a heterologous host to overcome the limitations of the limited supply obtained from natural producing organism. Two such examples are α-amylase of *Bacillus stearothermophilus* and cellulase of alkiliphilic *Bacillus BCE103*. Amylases have been used for many years to remove starch sizes from fabrics, a process termed as desizing. Originally, the source of amylase was plants and animals, which were later, replaced by bacteria. The first bacterial enzyme for desizing was α-amylase from *Bacillus subtilus*, which was commercialized in the early 1950s. Whereas a novel α-amylase naturally produced by *Bacillus stearothermophilus* is heat stable and active over a broad pH range. However, the attractive new α-amylase is produced at low levels by its natural producing organism (Sinclair, *et al.* 2004). Therefore, it is manipulated for the commercial use; the gene encoding the enzyme was cloned into a heterologous host *Bacillus licheniformis*.

Other enzyme cellulases prevent and remove fuzz and pills and provide color brightening of cellulose-based fabrics such as cotton. A novel cellulase, active under alkaline detergent conditions is an attractive alternative. As it is also produced at very low quantity by its naturally producing organism, the gene encoding the cellulase was cloned into a heterologous host, *Bacillus subtilus*, for high level expression and enzyme production. As a result, the desired enzymes were produced efficiently from heterologous hosts during fermentation in both the cases. In addition, the enzymes are secreted directly into the culture media, which simplifies the recovery of the respective enzymes.

Production of Enzymes from Gmms

Enzymes manufactured by GMMs have been used in the food industry for more than 15 years. Well known examples include the use of chymosin for cheese making and pectinases for fruit and beverage processing. Traditionally, cheese making required chymosin-containing rennet from calf stomach to provide the essential proteolytic activity for coagulation of milk proteins. However, chymosin preparations could have contaminants from animal source. In the early 1980s, Gist-brocades generated the hopes of producing chymosin from a microorganism using genetic engineering approach. The gene encoding calf stomach chymosin was cloned and expressed in an industrial strain of *Kluyveromyces lactis*, yeast that has been used for many years in the safe production of food ingredients. Chymosin produced through this genetically engineered yeast strain has the same chemical and biological properties as that from calf rennet. The chymosin preparation registered under the brand name Maxiren[®] has been commercially produced since 1988. Also, production of pectinase via genetic engineering approach focuses on economic enzyme production, enhanced enzyme purity and environment friendly production processes (Wu, *et al.*2002). It is possible to produce enzymes economically with a higher specificity and purity by the use of GMM. Some enzymes produced artificially from modified microorganisms are Glucoamylase, Polygalacturonase, Cellulase, Pullulanase, Alpha-amylase, Xylanase, Lipase, Alpha-acetodecarboxylase, Protease, Glucoamylase etc.

Recombinant Therapeutic Protein Production

Several proteins such as insulin, Interferons (IFNs) and Interleukins (ILs) are now produced by GMMs for therapeutic use. The traditional method of supplying these proteins to the patient requires purification of proteins from cells, tissues or organs of humans, cows or pigs. Because it was impractical to treat diabetes with human insulin cadaver sources, cow and pig insulin are somewhat different from human insulin, were substituted. The problem with obtaining the protein directly from animal sources included the limited supply and potential immunological responses. Limited supply translates into higher cost for the medication. Further concerns arose therapeutic proteins of animal origin may be contaminated with viruses or other toxic substances (Wolfenbarger, & Phifer, 2000). These problems can be avoided by producing the proteins in microorganisms. Human insulin, the first recombinant therapeutic protein approved by Food and Drug Administration (FDA) in 1982, was produced by genetically engineered *E.coli* containing the human insulin genes. Human growth hormone approved by FDA in 1985 was produced by a modified *E. coli* strain containing the native human growth hormone gene. Production of these therapeutic proteins in a fast growing and easily manipulated organism ensures sufficient supply, free of contamination, reduced cost, safe and consistent production.

REFERENCES

[1]   Bakan, B. Melcion, D. Richard, M.D. & Cahagnier, B. (2002). Fungal growth and Fusarium mycotoxin content in isogenic traditional maize and genetically modified maize grown in France and Spain. *Journal of Agricultural and Food Chemistry*, 50: 728–731.

[2]   Carpenter, J.E. (2001). Case studies in benefits and risks of agricultural biotechnology: Roundup Ready soybean and Bt field corn, National Center for Food and Agricultural Policy, Washington DC.

[3]   Conner, A.J. Glare, T.R. & Nap, J. P. (2003). The release of genetically modified crops into the environment. Part II. Overview of ecological risk assessment. *Plant Journal*, 33: 19–46.

[4]   Crailsheim, K. (1990). The protein balance of the honey-bee worker. *Apidologie*, 21: 417–429.

[5]   Dale, P.J. Clarke, B. & Fontes, E.M.G. (2002). Potential for the environmental impact of transgenic crops. *Nature Biotechnology*, 20: 567–574.

[6]   Gianessi, L.P. (2005). Economic and herbicide use impacts of glyphosate-resistant crops. *Pest Management Science*, 61: 241–245.

[7]   Glare, T.R. & O'Callaghan, M. (2000). Bacillus thuringiensis: Biology, Ecology and Safety, John Wiley and Sons Ltd, Chichester UK.

[8]   James, C. (2004). Preview: Global status of commercialized transgenic crops 2004, ISAAA Briefs No. 32. ISAAA, Ithaca NY.

[9]   James, C. (2005). Global status of commercialized biotech/GM crops: 2005, ISAAA Brief No. 34. International Service for the Acquisition of Agri-biotech Applications, Ithaca, NY.

[10]   Malone, L.A. & Pham-Delegue, M.H. (2001). Effects of transgene products on honey bees (Apis mellifera) and bumblebees (Bombus sp.). *Apidologie*, 32: 287–304.

[11]   Munkvold, G.P. Hellmich, R.L. & Rice, L.G. (1999). Comparison of fumonisin concentrations in kernels of transgenic Bt-maize hybrids and nontransgenic hybrids. *Plant Disease*, 83: 130–138.

[12] Munkvold, G.P. Hellmich, R.L. & Showers, W.B. (1997). Reduced Fusarium ear rot and symptomless infection in kernels of maize genetically engineered for European corn borer resistance. *Phytopathology*, 87: 1071–1077.

[13] Romeis, J. Meissle, M. & Bigler, F. (2006). Transgenic crops expressing *Bacillus thuringiensis* toxins and biological control. *Nature Biotechnology*, 24: 63–71.

[14] Sinclair, T.R. Purcell, L.C. & Sneller, C.H. (2004). Crop transformation and the challenge to increase yield potential. *Trends in Plant Science*, 9: 70–75.

[15] Snow, A.A. Andersen, B. & Jørgensen, R.B. (1999). Costs of transgenic herbicide resistance introgressed from Brassica napus into weedy *B. rapa*. *Molecular Ecology*, 8: 605–615.

[16] Snow, A.A. Andow, D.A. Gepts, P. Hallerman, E.M. Power, A. Tiedje, J.M. & Wolfenbarger, L.L. (2005). Genetically engineered organisms and the environment: Current status and recommendations. *Ecological Applications*, 15: 377–404.

[17] Snow, A.A. Pilson, D. Rieseberg, L.H. Paulsen, M.J. Pleskac, N. Reagon, M.R. Wolf, D.E. & Selbo, S.M. (2003). A Bt transgene reduces herbivory and enhances fecundity in wild sunflowers. *Ecological Applications*, 13: 279–286.

[18] Wang, W.X. Vinocur, B. & Altman, A. (2003). Plant responses to drought, salinity and extreme temperatures: towards genetic engineering for stress tolerance. *Planta*, 218: 1–14.

[19] Wolfenbarger, L.L. & Phifer, P. (2000). The ecological risks and benefits of genetically engineered plants. *Science*, 290: 2088–2093.

[20] Wu, K. Li, W. Feng, H. & Guo, Y. (2002). Seasonal abundance of the mirids, Lygus lucorum and Adelphocoris spp. (Hemiptera: Miridae) on Bt-cotton in northern China. *Crop Protection*, 21: 997–1002.

[21] Yamaguchi, T. & Blumwald, E. (2005). Developing salt-tolerant crop plants: challenges and opportunities. *Trends in Plant Science*, 10: 615–620.

In: Recent Trends in Biotechnology and Microbiology
Editors: Rajarshi Kumar Gaur et al., pp. 189-198

ISBN: 978-1-60876-666-6
2010 Nova Science Publishers, Inc.

Chapter 16

PATENTING OF LIFE FORMS: INDIAN SCENARIO

Anvita Parashar
Amity University Rajasthan, Jaipur

ABSTRACT

Today we are living in the age of Information Technology and spreading our wings to every aspect of the society. We exchange our information, as it is the human tendency to share their innovative thoughts with their near and dear ones but that proves fatal, in most of the cases, in respect of that original creation or property i.e. Intellectual Property (IP). IP laws are extremely important for the scientific development, growth and prosperity of any country. The international community realized this around the early 1990's and fallout of this realization came out in the form of Trade Related Aspects of Intellectual Property Rights (TRIPs) agreement in 1994 and its enforcement on 1st January 1995. A patent is an exclusive right granted to inventor or creator of a useful or improved article or a new process of making an article for a specified period of time.

India is a member of the World Trade Organization (WTO) and a signatory to the WTO agreement on TRIPs. Pursuant to the TRIPs agreement, India has amended its patent legislation which is governed by the Patent Act, 1970 on three occasions. The case of Diamond V. Chakrabarty in 1980 led to general trend of patenting inventions on living matter. In India, the position was made clear after the 2002 amendment to the Indian Patents Act. The amended act stated that life forms can be patented provided they satisfy the other requirements. The improvements in the Indian patent regime have resulted in a significant up thrust in the promulgation and enforcement of patents in India. Now, India can boast of one of the best patent law regimes in the world. However, a lot remains to be achieved still. Improving IPR protection will be an important element to increase the attraction of private investors in India.

INTRODUCTION

Today, we are living in the age of information technology and spreading our wings to every aspect of the society. We exchange our information (ideas, techniques, process, and product) to a target group but we never come to know this thing that someone is also targeting our information. So we are prone to unintentional leakage of our precious ideas as our information passes through different portals. It is human tendency to share their innovative thoughts with their near & dear ones but that proves fatal, in most of the cases, in respect of that original creation or property i.e. intellectual property (IP). Most of the people act indiscreetly & ignorantly which results in jeopardizing the chance of saving their intellectual property i.e. novelty of ideas as they come under public domain & lose their chance to be protected by the INTELLECTUAL PROPERTY RIGHTS (IPR). The term Intellectual Property reflects the idea that its subject matter is the product of the mind or the intellect. Intellectual property rights are legal property rights over creations of the mind, both artistic and commercial, and the corresponding fields of law. Under intellectual property law, owners are granted certain exclusive rights to a variety of intangible assets, such as musical, literary and artistic works; ideas, discoveries and inventions; and words, phrases, symbols, and designs. IP is the foundation of knowledge-based economy. The majority of intellectual property rights provide creators of original works economic incentive to develop and share ideas through a form of temporary monopoly (Hahn, 2005).

IPRs are needed as it helps to protect investment of time, money, effort & such other resources of the inventor or creator. IP provides a pool of information to the general public since all forms of IP are published in journals & magazines except in case of trade secrets. It provides a mechanism of handling infringement, piracy and unauthorized use, not only this; it also encourages industrial development and technological advancement which leads to overall economic development of the country (Lessig, 2004). IPR are bundle of rights i.e. it includes various independent rights. Following are the various independent rights for which IPR collectively provides protection:

1) Patent Right (Patent Act, 1971 & Patent Rules, 2000)
2) Industrial Design (Design Act)
3) Trademarks (Trademarks Act)
4) Copyright (Copyrights Act)
5) Geographical Indication (Geographical Indication Of Goods Act)
6) Trade secrets (Common Law)
7) Circuit Layout Design (Semiconductor Layout Design Act)

Amongst the IP rights the one which is of our concern is the area of patents. A patent is an exclusive right granted to inventor or creator of a useful or improved article or a new process of making an article for a specified period of time. After the expiry of the duration the invention becomes part of public domain i.e. everyone can use it. So Patent means monopoly rights of inventor in respect of an invention. The word *patent* originates from the Latin *patere*, which means "to lay open" (i.e., to make available for public inspection) and more directly as a shortened version of the term *patent*, which originally denoted an open public reading royal decree granting exclusive rights to a person. Patents are a means of protecting inventions developed by firms, institutions or individuals and as such they may be interpreted as indicators of invention. Before an invention can become an innovation, further entrepreneurial efforts are required to develop, manufacture and market it. Patent indicators convey information on the output and processes of inventive activities.

CHARACTERISTICS OF THE PATENT

- ❖ Novelty: It cannot have existed previously.
- ❖ Usefulness: It must have some utility.
- ❖ Inventive: Only inventions can be patented. Invention means a new product or a new process involving an inventive step and capable of industrial application. Invention includes within its scope any new and useful improvements of any manner of manufacture, article or substance.
- ❖ Originality of inventorship: Only those who have contributed intellectually may be on the patent application. It is not the same as authorship.
- ❖ Full disclosure: All secrets of the invention must be disclosed. It is the opposite of trade secrets. A patent is a bargain with the state in which the government says, "You teach the country how to use this invention properly and in return, we'll give you a limited monopoly on the use of it for 20 years."

Ownership of the Patent

An application for a patent may be made by the inventor, either alone or jointly with another, or his/their assignee, legal representative of deceased inventor or assignee are entitled to apply. For e.g. If a person invents a new product or process & unfortunately soon after that he dies then his legal heirs can or any person authorized by him before his death can apply for patent.

Geographical Limits of the Patent

Patents are national. Patent is granted for a specific invention in a particular country in which an application is made for the same cause. There is no international patent as such though it has acquired an international character. For e.g. a patent granted in India is valid only for India and not in the USA. However, a patent granted in the EPO is valid in all the contracting states recognized by European Patent Organization. The protection so granted in a country/region not only identifies the rights of the creator/inventor or his assignees, but also enables the right holder to enforce his rights against infringers (Craig, 2005).

Term of the Patent

Term of every patent is usually 20 years from the date of filling of patent application & date of patent is the date on which the application for patent is filed, irrespective of the fact whether it is filed with provisional or complete application. To keep the patent in force renewal fee is to be paid every year.

Effect of the Patent

A patent is not a right to practice or use the invention but it is a negative right. When one owns a patent, this does not give him the right to practice his own invention, it only gives him the right to prohibit others from making, using, and selling, offering for sale, or importing the patented invention for the term of the patent, which is usually 20 years from the filing date subject to the payment of maintenance fees. A patent deprives the public of nothing that it freely enjoyed prior to the grant of the patent. A patent adds to the sum of human knowledge and encourages further interaction.

Laws Governing the Patent

The grant and enforcement of patents are governed by national laws and also by international treaties, where those treaties have been given effect in national laws. There is a trend towards global harmonization of patent laws, with the World Trade Organization (WTO) being particularly active in this area. The Trade Related Aspects of Intellectual Property Rights (TRIPs) Agreement has been largely successful in providing a forum for nations to agree on an aligned set of patent laws. Conformity with the TRIPs agreement is a requirement of admission to the WTO and so compliance is seen by many nations as important. Moreover, several international agreements, treaties & conventions exist to monitor that the inventor/creator are not denied of his/her rights like European Economic Community Treaty (EEC), Patent co-operation Treaty (PCT), European Patent Conventions and Protocols (EPC), Community Patent Convention and Protocols (CPC) resulting in a common patent office for granting common patents applicable to the member countries. Under the PCT, a person can file a single application to seek protection in all of the contracting parties to the PCT. The process of a PCT application can be divided in two phases: international and national. The international phase starts with the rights holder filing a PCT application in an eligible receiving office or the International Bureau in Geneva. Thereafter, there are steps for international search, publication and preliminary examination. Subsequent to this is the national phase, in which the rights holder has to pay fees and prosecute an application in all the national or regional offices where protection is sought. The procedure for prosecution of a patent application then becomes dependant upon the laws of each contracting party (Stix, 2006).

In India, the patent legislation is governed by the Patent Act, 1970 which was an outcome of various previously existing patent legislations as outlined in the table below.

India is a member of the WTO and a signatory to the WTO Agreement on TRIPs. As a developing country, India was given until January 1, 2005 to effect full implementation of its obligations under TRIPS. Pursuant to the TRIPs Agreement, India has amended its patent legislation in the years 1999, 2002 and 2005. The amendments were made in order to bring the patent law in India into compliance with the TRIPs Agreement and to introduce patents for drugs, medicines and food products; to remove transitional provision to exclusive marketing rights, to rationalize and reduce the timeline for processing patent applications.

Table 1. History of Indian Patent System.

1856	The act vi of 1856 on protection of inventions based on the british patent law of 1852. certain exclusive privileges granted to inventors of new manufacturers for a period of 14 years.
1859	The act modified as act xv; patent monopolies called exclusive privileges (making. selling and using inventions in india and authorizing others to do so for 14 years from date of filing specification).
1872	The patents & designs protection act.
1883	The protection of inventions act.
1888	Consolidated as the inventions & designs act.
1911	The indian patents & designs act.
1972	The patents act (act 39 of 1970) came into force on 20th april 1972.
1999	On march 26, 1999 patents (amendment) act, (1999) came into force from 01-01-1995.
2002	The patents (amendment) act 2002 came into force from 2oth may 2003
2005	The patents (amendment) act 2005 effective from 1st january 2005

Patent Application

The Indian Patent Office has its head office at Kolkata, which has three branch offices located at Mumbai, Chennai and Delhi. The Controller General heads the Patent Office and each branch has a Controller as its head. In case of an Indian applicant, the patent application must be filed at the patent office under whose jurisdiction the applicant has his place of work, or place of residence or place where he conduct business from. In case of foreign applicant/s, the jurisdiction in which the patent application is filed would be based on the address for services of the applicant's agent. For e.g. if the address for services for foreign applicant is based at Bangalore, the patent application must be filed at the Chennai Patent Office. The official patent office site is http://patentoffice.nic.in/ipr/patent/patents.htm. The salient points of the patent process which needs to be meticulously followed in India are:

1) Patent search.
2) Filing the provisional application with the Patent Office with provisional specifications as soon as the nature of the invention is conceived and demonstrated.
3) Filing the complete specification as soon as the invention, as a whole, has been successfully and reproducibly worked/demonstrated at the lab.
4) Informing the patent office at the earliest about any case of possible infringement of patent rights, with all available documentary evidence for filing legal notice or initiating legal action.
5) Do not refer any technology to industry before a specification-complete application for patent in respect of the said invention has been filed at the Patent Office. Such reference, if needed, must be on the basis of an agreement for secrecy to be executed by the parties.
6) Do not publish/read a paper or display the invention or the design anywhere in the world before filing of a patent.

7) Do not transfer a technology to industry without identifying it with the patent application number/patent number in case an application for patent for the invention is pending or patent has been secured.

8) Do not enter into joint research/development work agreements which involve taking patents/designs/trade marks/copyright without checking with the patent office.

Biological Patent

A biological patent is a patent relating to an invention or discovery in biology/life forms. Biological inventions have been around for a long time. Sir Louis Pasteur got a US patent on yeast culture in 1873. The 1970s marked the first time when scientists patented methods on their biotechnological inventions with recombinant DNA. It wasn't until 1980 that patents for whole-scale living organisms was permitted. In U.S.A however the case of Diamond vs Chakrabarty opened new vistas of bringing forms of life, specifically microorganisms under the preview of patenting; in the case Supreme Court of the United States considered the question of whether a microorganism is patentable subject matter under the United States patent laws. Chakrabarty, a microbiologist, sought to patent a genetically engineered bacterium which degrades crude oil; a characteristic which makes it extremely valuable for controlling oil spills. The patent examiner denied Chakrabarty's patent claim for the bacteria itself, but allowed his claims for products and processes involving the bacteria. The examiner denied the bacteria patent, finding that a micro-organism is a product of nature which, is a living thing and thus cannot be patented. However when the case reached the Supreme Court, the Court held that a live, human-made micro-organism is patentable subject matter as a manufacture or composition of matter. Since legal changes have occurred starting in 1980, there has been a general trend of patenting inventions on living matter (Heller & Eisenberg, 1998). Among all countries and international bodies granting patents, the U.S. patent system is probably the most important and accommodating for biotechnology inventions. The U.S. places few if any restrictions on biotechnology and pharmaceutical inventions which may be patented. Most European countries do not issue utility patents for human and animal therapeutics and diagnostics and have avoided granting patents for modified organisms. A number of countries worldwide do not allow patenting of human pharmaceuticals. Obtaining U.S. patent protection is likely to be pursued for nearly all potentially valuable biotechnology inventions (Shiva, 2005). In India, the position was made clear after amendments to the Indian Patents Act.

AMENDMENTS IN INDIAN PATENT REGIME

Section 3 (d) of the Patent Act categorizes new forms of known substances as non-patentable unless the new form exhibits enhanced efficacy of that known substance. The enhanced efficacy must exhibit significant differences in properties with regard to efficacy. Thus the dividing line between being an eligible subject matter to be considered for grant of a patent and not being patent eligible is the efficacy.

Under Section 3 (h) of the act a method of agriculture and horticulture cannot be patented.

Section 3 (i) of the act states that any process for the medicinal, surgical, curative, prophylactic, diagnostic, therapeutic or other treatment of human being or any process for a similar treatment of animals to render them free of disease or to increase their economic value or that of their products in non-patentable.

Before the amendment, the unamended Section 3(j) of the Act stated that plants and animals in whole or in part thereof including seeds, varieties and essentially biological process for the production of plants and animals are excluded. However in India after the amendment micro-organisms can be patented provided they satisfy the other requirements (Mehra, 1999).

Strict biosafety norms have to be followed while handling microorganisms so as to ensure that these don't land into wrong and/or technically incompetent hands. Our patent laws for protecting microorganisms have spelled out requirements for the deposition of such life forms in any of the International Depositary Authorities (IDA) under the Budapest Treaty on or before filing the application. For the purpose of life forms Microbial Type Culture Collection and Gene Bank (MTCC), Chandigarh is an internationally recognized depository institution in India.

DIFFERENT ASPECTS OF PATENTING LIFE FORMS IN INDIA

Patenting of life forms has many dimensions. India is a storehouse of biological resources. In a world that patents DNA, India must follow current trends to advance its research and development. In recent years the rise in investment in biotech oriented industries is poised to take India to a different level in the world market. However, western nations are at an advantage in research given their better technological and financial resources. As scientific research advances, more patents of human DNA and cell line will emerge, as will many fundamental questions on human life for which there are no correct answers. Patenting can help further scientific development by making research public. Indians also risk losing monopoly over scientific advancements involving indigenous people, plants, and animals (Mashelkar, 2006). Hence, they must be able to apply for patents in their own country, enabling them to have monopoly and financial rights over their own scientific findings.

Another dimension is the extent of private ownership that could be extended to life forms. Micro-organism when genetically modified falls in the category of invention because of human input. Genetically modified micro-organism may perform any number of activities. If a researcher is able to research upon a particular activity and he is allowed patenting of his genetically modified micro-organism this will result in blocking of further research on that micro-organism. Since only inventions are qualified for patenting, naturally found micro-organisms, DNA structure, genes, blood cells, etc. can be excluded from patent protection. Clearly, we must re-examine the need to grant patents on life forms anywhere in the world. Nations can also exclude certain inventions in biotechnology by relying on the exclusion provision available under the TRIPS Agreement which permits the state parties to exclude certain inventions which are injurious to health and environment of human and animals

(Sharma, 2004). Using this exception a member state can exclude terminator type technologies from patent protection.

Looking with closer analysis, amendment of the Indian Patent laws have jeopardized our food security and hence our national security. Firstly, it allows patents on seeds and plants through sections 3 (i) and 3 (j), as we saw above. Patents are monopolies and exclusive rights which prevent farmers from saving seeds; and seed companies from producing seeds. Patents on seeds transform seed saving into an "intellectual property crime". Secondly, when combined with the product patents of the 3rd Amendment, Patents on Life Forms in the 2nd Amendment can mean absolute monopoly. Patents on seeds are a necessary aspect of the corporate deployment of genetically modified seeds and crops. Patent protection implies the exclusion of farmers' right over the resources having these genes and characteristics. This will undermine the very foundations of agriculture. Thus a company can introduce traits through genetic engineering and then claim monopoly on the trait even in traditional varieties through a product patent. A product patent in effect says that it does not matter how a property was created, came into existence, whether a result of evolution, or farmers breeding, or genetic pollution is patent infringement and theft.

One of the most significant impediments to the effective enforcement of patent rights in India is the acute lack of awareness of patent basics. Patent examiners, patent attorneys, scientists and technologists working in the area of biotechnology are not really familiar with the patenting of microorganisms and microbiological inventions. The technical complexity involved in the patenting of life forms need to be understood by such professionals. Efforts must be made to understand the legal, social, scientific, clinical and psychological effects of patenting genetic material. On one hand there is a need to create awareness among scientists; on the other hand there is an urgent need for training patent examiners and attorneys to handle applications in the area. Indian companies, inventors, investors and physicians venturing into the biotech sector must be well informed and aware of domestic and international laws as they seek to join the biotechnology competition.

ETHICAL AND MORAL CONSIDERATIONS OF PATENTING LIFE FORMS

A range of patents are unethical; they destroy livelihoods, contravene basic human rights and dignity, compromise healthcare, impede medical and scientific research, create excessive suffering in animals or are otherwise contrary to public order and morality. Some of the ethical and moral considerations associated with patenting life forms are as follows:

➢ Patents in science promote secrecy and hinder the exchange of information. By patenting products of research, the free flow of ideas and information necessary for cooperative scientific efforts is reduced. Scientific materials required for research will become more expensive and difficult to purchase if one corporation owns the rights to the material.

➢ Patents make important products more expensive and less accessible. Microorganisms, plants, animals and even the genes of indigenous people have been patented for the production of pharmaceuticals and other products. Requiring developing nations to pay royalties to the wealthy industrial nations for products derived from their own natural resources and innovation is robbery. The biotech industry claims that patents are necessary so that innovative, life-saving technologies will be developed. In actuality, patents enable companies to create a monopoly on a product, permitting artificially high pricing [10].

➢ Patents promote unsustainable and inequitable agricultural policies. A disastrous decline in genetic diversity could be the result of patenting of crop species. The genetic diversity inherent in living systems makes patent claims difficult to defend. The development of genetically uniform organisms would make it easier for corporations to maintain their patent claims. Biotech companies holding broad spectrum patents on food crops will lure farmers to grow modified varieties with promises of greater yields and disease resistance. However, numerous examples worldwide show the "improved" crops have failed to hold up to corporate promises, and led to the loss of the rich diversity of traditional crop varieties.

➢ Patents on life forms are morally objectionable to many. Patenting organisms and their DNA promotes the concept that life is a commodity and the view that living beings are "gene machines" to be exploited for profit. If it is possible to consider a modified animal an invention, are patents and marketing of human reproductive cells far behind? Patents derived from concepts of individual innovation and ownership may be foreign to cultures which emphasize the sharing of community resources and the free exchange of seeds and knowledge.

The moral and ethical questions raised are of fundamental importance. Should we tinker with the basic building blocks of the planetary environment and of life itself? What are the risks and what may be the consequences to our way of living? Do we as human beings have the right to meddle with a set-up which took so long to produce us, when we have only existed for such a comparatively short time? Are we risking the survival of humankind at a time of exponentially accelerating scientific and technical knowledge, when our human relationships, basic human rights, social care, human equality, freedom and independence are so inadequate over much of the planet, leave so much to be desired? Quite apart from the moral and ethical considerations involved in creating a new life-form, the public health risks are extremely high when tinkering with and changing in the twinkling of an eye and on a massive scale, life-forms which have taken millions of years to evolve slowly by trial and error and by eliminating inadequate or mistaken change. Unpredictable are the resulting direct and indirect effects on the human being, which is the most complicated organism ever produced and which we do not fully understand. These issues on patents and life forms will not disappear. They will have to be addressed the Patent debate is not over, it has just begun. And in changing the distorted, unjust, illegitimate "intellectual property" regime of W.T.O. local and national actions will be as relevant as international negotiations.

REFERENCES

Hahn, R. W. (2005). Intellectual Property Rights in Frontier Industries: *Software and Biotechnology*. AEI Press.

Lessig, Lawrence. (2004). Free Culture: How Big Media Uses Technology and the Law to Lock Down Culture and Control Creativity. New York, Penguin Press.

Craig, Fellenstein. (2005). The Inventor's Guide to Trademarks and Patents. Prentice Hall PTR.

Stix, Gary. (2006). Owning the Stuff of Life. Scientific America, Volume 294: Issue 2.

Heller, M.A., & Eisenberg, R.S. (1998). Can Patents Deter Innovation? The Anticommons in Biomedical Research. *Science*, 280, 698-701.

Shiva, V. (2005). WTO, Patents on Lifeforms and Amendments in India's Patent Law. Z-Space The Spirit of Resistance Lives. Available from www.zmag.org/zspace/commentaries/2243

Mehra, K. L. (1999). Understanding the Patent Law, its Flaws. Indian Express Newspapers. Available from: ww.indianexpress.com/ie/daily/19990222/ibu22068.html

Mashelkar, R. A. (2006). Report of the Technical Expert Group on Patent Law Issues.

Sharma, D. (2004). Deaths knell for low cost medicines. India Together. Available from: www.indiatogether.org/2004/dec/dsh-3rdamend.htm

In: Recent Trends in Biotechnology and Microbiology
Editors: Rajarshi Kumar Gaur et al., pp. 199-207

ISBN: 978-1-60876-666-6
2010 Nova Science Publishers, Inc.

Chapter 17

IPR AND AGRI-BIOTECH MANAGEMENT IN INDIA – THE BIG PICTURE

Ashish Kumar Sharma, Sushil Kumar Rai and Anjali Sharma

Department of Commerce, Faculty of Arts, Science & Commerce,
Mody Institute of Technology & Science, Lakshmangarh, Distt.-Sikar,
Rajasthan, INDIA PIN-332 311

ABSTRACT

The biotechnology industry is one of the most research and development intensive and capital-focussed industries in the world. As such, its success relies on a robust intellectual property system as well as a strong set of competition laws. The management of biological resources has been an increasingly contentious subject at the national and international levels. This is linked in large part to the progressive recognition of new economic opportunities arising from the use of biodiversity. As a result, international legal frameworks for the management of biological resource have had to increasingly take into account not only the needs of biodiversity conservation but also concerns about its potential for economic use and its contribution to the process of economic development. This has important repercussions from a legal perspective because the new products developed by the biotechnology industry can often easily be copied once they have been put on the market. As a result, the biotechnology industry has strongly argued for the introduction of Intellectual Property Rights (IPR) over genetically modified organisms. This paper seeks to analyze the impact of the international legal framework for the promotion of IPR on India's legal regime concerning the control over biological resources and inventions derived from biological resources. It focuses in particular on three acts and legislative amendments adopted in recent years and their organic relationship within the overall domestic legal framework. The paper analyses these enactments in the context of the move towards the control of biological resources and derived products through IPR. This has ramifications not only for control over biological resources and derived products but also more generally on the management of agriculture in India and other developing countries and the realization of food security and the human right to food at the individual level.

"These varieties have 50 percent higher yields, mature 30 to 50 days earlier, are substantially richer in protein; are far more disease and drought tolerant, resist insect pests and can even out-compete weeds. And they will be especially useful because they can be grown without fertilizer or herbicides, which many poor farmers can't afford anyway. This initiative shows the enormous potential of biotech to improve food security in Africa, Asia and Latin America." [1]

ROLE OF INTELLECTUAL PROPERTY RIGHTS (IPR) IN THE BIOTECHNOLOGY INDUSTRY

The biotechnology industry is very much dependent upon strong intellectual property protection. Biotech companies rely heavily on private investments for financing, and patents (one of the most important and widely used intellectual properties) are among the most important benchmarks of progress in developing a new firm's product line. Empirical evidence suggests that patents are an important means for protecting the value of biotech firms. Economist Joshua Lerner has found that not only do patents help biotech firms attract venture financing, but also that the valuation of a start-up biotech firm is directly proportional to the number and scope of the firm's patents. In another recent economic analysis, specific values were attributed to biotechnology patents as calculated from a function dependent on the content of the patents. Individual patents in the core areas of biotechnology development were shown in this study to be valued between $13 and $21 million on an average. This same analysis also revealed that a biotechnology patent yields significant economic value to rival firms of the patent holder due to the public knowledge spill-over from the patent's disclosure.

Biotech firms recognize the value of patents in their industry and survey evidence has tended to show that the biotechnology industry places a greater emphasis on seeking patent protection than do many other industries. In a 1989 study, firms in over 100 industrial disciplines ranked patents as the least important of several available strategies for competing in new-product markets. These firms placed more emphasis on the effectiveness of trade secrecy, early entry, and customer service in emerging product sectors. The same study revealed, however, that firms in the chemical and pharmaceutical disciplines (into which most biotech firms were categorised at the time) ranked patents as one of the most effective means for effecting competition. In a similar study undertaken in 1986, empirical data suggested that not only did the chemical and pharmaceutical industries place more emphasis on patents, but a relatively large portion of the innovations in those industries would never have succeeded in the market without patent protection (as compared to innovations in ten other industries). The 1986 study also showed that in those industries (including chemical and pharmaceutical) where patents were regarded as a relatively important means for competition, patents were more likely to be sought for patentable inventions than in those industries that did not regard patents as relatively important.

The results of these mid-1980's studies were re-examined in a 1999 survey, which delineated biotech firms from the chemical and pharmaceutical industries. This survey suggested that in industries other than biotechnology – chemical and pharmaceutical – to

---

[1] Mark Malloch Brown, United Nations Development Program administrator, quoted in "Report Cites Benefits of Biotechnology for Developing Countries," Environment News Service, 11 July 2001

which patents remain very important, the emphasis on secrecy and non-disclosure has generally increased since 1987. The importance of patents to the biotechnology industry, however, remains high. This can be seen because biotech firms have continued to file more and more patent applications even though the cost of filing and presenting biotechnology patents is higher and it takes longer than inventions in some other fields because long delays in the Patent and Trademark Office are common for biotechnology patents.

IPR AND BIOLOGICAL RESOURCES IN INDIA IN THE TRIPS ERA

The impact of the domestic legal regime on IPR must be understood in the context of the rapid changes observed at the international level over the past decades. At the international law level, the distribution of IPR over biological resources has been a long-standing concern. One of the basic principles of international law since decolonization has been the permanent sovereignty of states over their natural resources and, over time, the conservation and management of biological and genetic resources has evolved as a "common concern of humankind" irrespective of the nation to which they belong to.

Recent developments in India are interesting because India is one of the few countries with significant biological resources, the potential to develop its own biotechnology industry and strong local knowledge bases concerning the use of its biological resources. India is also one of the few countries that, while not rejecting the western patent model, decided as an independent country to tailor the system to make sure that it would fulfil a number of socio-economic goals in keeping with the overwhelming need to put the tackling of poverty before legal mechanisms providing protection to individual inventors.

The legal status of plant genetic resources constitutes an interesting case study. In general, the call for the establishment of the principle of permanent sovereignty over natural resources first came from newly decolonized countries. However, with regard to plant genetic resources, developing countries argued in favour of the concept of "common heritage of humankind" which implies that no sovereign rights can be imposed. This was opposed by some developed countries but the international community adopted the International Undertaking, which recognized plant genetic resources as a common heritage of humankind. The rationale was that states should collaborate in the management of plant genetic resources because improved flow of germplasm and related knowledge between countries would contribute to the broader goal of enhancing food security in developing countries.

In fact, the International Undertaking reflected the existing international system for agricultural management. India had, like many other developing countries, substantially benefited from the principle of free sharing of knowledge and resources. Green Revolution varieties, which contributed to significant yield improvements in some regions of India for a number of years, were the product of international collaborative research and the result of the exchange of crop varieties across countries and regions of the world. This international collaborative system was formalized in the early 1970's with the setting up of the Consultative Group for International Agricultural Research (CGIAR), a coordinating institutional structure for a series of international agricultural research centres spread around the world.

Following establishment of the international institutional mechanisms such as the Convention on Biological Diversity (CBD) and the WTO, and further, signing of International Treaty on Plant Genetic Resources for Food and Agriculture (ITPGRFA), the growing importance and the global scope of IPR in biotechnology are well realised and recognised. The IPR, after a long debate, is now recognised as an asset and means of rewarding and harvesting the fruit of biotech research and development. Recognition of IPR provides an effective means of protecting and rewarding innovators. This acts as a catalyst in technological and economic development. The essence of regulation of IPR by law is to balance private and public interests. At the same time, equitable benefit sharing is, although, agreed upon under the CBD, is yet to be realised in effective terms.

THE THREE IPR ACTS IN INDIA W. R. T. BIOTECHNOLOGY

In India, the three legislative instruments, viz. The Biological Diversity Act (2002), The Plant Variety Act (2001) and The Patents (Amendment) Act (2002), make up as a whole new legal regime for the use of biological resources and related knowledge. By collectively increasing the attractiveness of the economic exploitation of biological resources and related knowledge, they have resulted in two main consequences. One is their significant influence on conservation policies in India. This is important because while the Biodiversity Act adopts the twin goals of conservation and sustainable use, neither the Patents Act nor the Plant Variety Act are concerned with biodiversity conservation. This makes the striking of a balance between economic use and conservation difficult to achieve without specific coordination between the acts at the implementation level. The second main consequence of the approach adopted by the three instruments is in the field of property rights. The combined impact of the three enactments is an implicit (re)distribution of property rights. This has been controversial because of the socioeconomic implications and the need to find new ways to articulate private property rights and sovereign rights in the field of biological resources.

The Biodiversity Act clearly reflects the trends of the international level. It seeks at the same time to promote sovereign and private appropriation of biological resources and related knowledge. Among the consequences that this will have from the perspective of international collaboration on agricultural research is that India as a country will be less willing to share its resources freely with the CGIAR Centres. Since most countries are likely to adopt the same attitude, this will result in diminishing supplies of germplasm to international seed banks. This will be damaging to countries like India that have in the past significantly benefited from the international collaborative system. In fact, this is likely to go against the interests of most developing countries since most of them, including India, are highly dependent on genetic resources from other regions for their main staples.

In turn, the Biodiversity Act seeks to promote private appropriation through intellectual property rights. This has the potential to be beneficial to all entities or individuals around the world who make any new biodiversity-based product that fulfils the criteria for protection through intellectual property rights. However, there is no guarantee that this arrangement will be of most benefit to Indian individuals or entities despite the potential for the development of a domestic agro-biotechnology industry.

The rapid adoption of diverse systems of property rights to foster appropriation of biological resources and related knowledge has significant consequences for individuals or entities that cannot benefit from the new system in place. Two main examples illustrate this trend. *Firstly,* one of the characteristics of patents is to provide near monopoly rights that allocate all the benefits of a given invention to one entity or individual. In the case of agro-biotechnology, it is often the case that the final product, which can be protected by a patent such as a transgenic seed, is the product of diverse types of research and knowledge by different people in different places and at different times. One of the most frequent examples is the case where the local knowledge of a farmer or local community is used as a basis by researchers in the formal sector to develop a new transgenic product. Since the local variety cannot be protected through intellectual property rights because it does not fulfil the necessary requirements, no protection is offered by the legal system, which does not offer any alternative forms of protection at present. In the current scenario, the only thing that is offered to individuals or groups who have contributed to the development of a product protected by IPR, is "benefit-sharing." Benefit sharing as proposed in the acts analyzed above constitutes a form of compensation for the absence of property rights. In other words, a local farmer can, for instance, be granted a sum of money as compensation for his/her contribution to a patented invention but s/he cannot claim property rights over knowledge. In a historical perspective what has happened over the past two decades is that agricultural management has rapidly moved from a system where no one could claim any intellectual property rights in agriculture to a system where some can claim very effective rights (a commercial seed company for instance) and some do not get any property rights (traditional knowledge holders for instance).

Secondly, problems related to the development of the new property rights system also take the form of bio-piracy across countries. This is partly linked to the different systems through which countries judge "novelty" in the patents system and partly linked to the different levels of IPR protection in different countries. In the case of the former, the kinds of problems that can surface are well illustrated by a patent taken in the United States on some supposedly new healing properties of turmeric. The patent was granted for properties that were well known in India but the US patent system does not specifically force patent examiners to take into account knowledge from other countries. As a result, the Indian government had to contest the patent in the United States to have it revoked. The latter case is illustrated by a number of patents taken in the United States or in Europe on agricultural processes that were not patentable in India. Bio-piracy is problematic for a number of reasons. While the worst cases—such as the turmeric patent where the patent should not be granted in any jurisdiction because the conditions for patentability are not fulfilled—can be eliminated by improving access to traditional knowledge, developing countries will find it difficult to stop the appropriation of their knowledge in other countries as a whole. This is because the TRIPS agreement only imposes minimum levels of IPR protection. Member states can go further than the minimum levels. As a result, inventions that may not be patentable in India may be patentable in the United States or Europe. Since patents are territorial rights, the only thing that a country like India can do is to restrict patentability to the extent possible under TRIPS in its own jurisdiction. This still makes it possible for knowledge to be patentable in other jurisdictions.

The appropriation of biological resources and related knowledge in India must also be seen in the context of the historical debate that led to the adoption of the Patents Act, 1970 and the socioeconomic concerns that were introduced at that point. The Patents Act specifically sought to accept patents as a useful tool to reward inventiveness while recognizing that the system had to be carefully bounded to avoid undesirable social outcomes. This led to the adoption of provisions to ensure that patents rights would not be used in a manner detrimental to the public at large. The act imposed, for instance, restrictions meant to avoid the over-commercialization of sectors that were of vital importance for meeting basic needs, such as food and health. It prohibited the patentability of all methods of agriculture and horticulture or processes for the medicinal, surgical or other treatment of human beings. The act also limited the term of process patents for substances intended for use as food, medicine or drug, the term was seven years while the normal term of the patent was fourteen years. Further restrictions were imposed on the rights of the patent holder including stringent provisions for compulsory licensing and for licenses of right.

The new regime adopted is surprising not only because it dismantles the restrictions put in place for socioeconomic reasons but also because it does not seem to provide an integrated response to existing challenges. The most significant element is probably the fact that the question of the relationship between the patent system and sustainable biodiversity management has been addressed neither in the Biodiversity Act nor in the Patents Amendment Act. This includes, for instance, issues concerning the possibility offered to the National Biodiversity Authority to impose the sharing of intellectual property rights as a form of benefit-sharing, something which is not provided for in the Patents Act. This is only partly surprising since the same problem exists at the international level between the biodiversity convention and the TRIPS agreement and since the act and the Amendments were drafted separately in two different ministries.

Recent developments tend to indicate that a number of problems should be addressed at both the national and international levels. The nature of the current property rights regime that on the surface puts power in the hands of state by reaffirming sovereign rights over biological resources but in effect removes more power from their control by insisting on the increasing scope of private property rights must be addressed concurrently at the national and international levels. While for developed countries, the solution may lie in ever stronger patent rights, this cannot constitute a solution for countries like India, which need to take into account basic needs such as food while devising policies on biodiversity management. In fact, one of the most important tasks facing developing countries in years to come will be to "diversify" the property rights regime in the field of biological resource management. This implies that they should take steps to ensure that the rights of holders of knowledge, which are not easily protected under the existing legal regime such as traditional knowledge, are provided specific protection under national law. This constitutes one of the few avenues opened to India in the current environment to strengthen the legal framework in this area in favour of the majority of its citizens. It will also be necessary to address issues concerning the relationship between property rights and human rights, including the questions of restrictions on the patentability of life forms that should be imposed, as well as more specific concerns about the impacts of the introduction of patents in biological resource management on the realization of the human right to food, right to health and the right to a clean environment.

CONCLUSION

The influence of the TRIPS agreement over recent legislative activity is a fact that assumes more significance because its impacts go beyond the strict field of intellectual property. This is visible insofar as some of the changes imposed by TRIPS directly impact on environmental management and environmental laws while at the same time fostering a property rights regime that has the potential to have negative impacts on the management of biological resources. Importantly, this has largely been a one-way route. Environmental law and environmental concerns, domestic or international, in spite of their considerable stake in the shape of IPR regimes have had little impact on the development of the IPR regime.

India's reaction to TRIPS must be understood in this larger context. The government's initial resistance to the inclusion of intellectual property in the context of trade negotiations during the Uruguay Round gradually dissipated until it finally changed its mind. Though the first set of changes required by TRIPS was rejected in parliament, there has been a progressive softening of the opposition at a political level. Importantly, the current government headed by the Congress Party has been consistently in favour of WTO since coming to power. This U-turn from an earlier position where the party denied itself in terms of its opposition to external influences is substantially quailed by its continuing claims to represent indigenous interests as evidenced in its claims in favour of locally based indigenous development. The tension between these two irreconcilable objectives can partially account for the responses offered to the global IPR regime. In effect, the new legislative framework attempts to not upset the global legal order while simultaneously refusing to surrender the domestically significant currency of national interest. At the level of the legal regime, this has translated into a greater concentration of powers in the hands of the government at the national level accompanied by a surrendering of certain avenues to private sector interests.

Finally, it is significant that the shape of the legal framework reflects the extent of people's participation in the legislative process. The TRIPS agreement was widely criticized even before its adoption but widespread consultations within the country were never held before ratification. This missing participation was first reflected in parliament's rejection of the first proposed amendment to the Patents Act only half a year after the government had ratified the TRIPS agreement and committed the country to its implementation. Since then, some of the lessons of the lack of participation have been learnt, though in a yet unsatisfactory fashion. The Plant Variety Bill introduced in 1999 in parliament did not benefit from widespread participation by relevant actors but was significantly modified after a parliamentary committee conducted its own survey and decided the draft was not appropriate. In the context of the Biodiversity Act, a major consultative process has been taking place in the context of the development of a National Biodiversity Strategy and Action Plan but it has not had a direct impact on the act, which was drafted and adopted much before the completion of the strategy and action plan.

The interactions between intellectual property rights regimes and biodiversity management remain an evolving and unsettled issue at the international level. This notwithstanding individual countries like India must put in place legal frameworks for the management of biodiversity that make a coherent whole. While the existing national regime is insufficiently concerned with the overall coherence of the system put in place, it can be hoped that these shortcomings will be addressed at the level of implementation.

REFERENCES

[1]   Agreement Establishing the World Trade Organization, 15 April 1994. *Marrakech. International Legal Materials* 33: 1144.

[2]   Biodiversity Convention: Convention on Biological Diversity. 5 June 1992. Rio de Janeiro. *International Legal Materials* 31: 818.

[3]   Consultative Group on International Agricultural Research. 1999. CGIAR Center Statements on Genetic Resources, Intellectual Property Rights, and Biotechnology. Washington, DC: CGIAR.

[4]   Dhavan, Rajeev, & Maya Prabhu. (1995). Patent Monopolies and Free Trade: Basic Contradiction in Dunkel Draft. *Journal of the Indian Law Institute* 37: 194.

[5]   Eisenberg, S. Rebecca. (1987). Proprietary Rights and the Norms of Science in Biotechnology Research. *Yale Law Journal* 97 (2): 177–231.

[6]   International Undertaking: International Undertaking on Plant Genetic Resources, Res. 8/83, Report of the Conference of FAO, 22nd Sess., Rome 5–23 Nov. 1983, Doc. C83/REP.

[7]   Jodha, N. S. (1995). Common Property Resources and the Environmental Context. *Economic and Political Weekly* 30 (51): 3278–3283.

[8]   Kumar, Nagesh. (1998). India, Paris Convention and TRIPS. *Economic and Political Weekly* 33 (36–37): 2334.

[9]   Palacios, Flores Ximena. (1997). Contribution to the Estimation of Countries' Interdependence in the Area of Plant Genetic Resources. *Commission on Genetic Resources for Food and Agriculture*, Background Study Paper No. 7 Rev.1.

[10]   Paris Convention: Convention for the Protection of Industrial Property. 20 March 1883 (as revised). Paris.

[11]   Patents Act. 1970. Government of India, Act 39 of 1970.

[12]   Patents (Amendment) Act. 1999. Government of India. Gazette of India. 26 March 1999.

[13]   Plant Variety Act: Protection of Plant Varieties and Farmers' Rights Act, Government of India, 2001.

[14]   PGRFA Treaty: International Treaty on Plant Genetic Resources for Food and Agriculture, Rome, 3 Nov. 2001.

[15]   Report of the Joint Committee. 2000. Joint Committee on the Protection of Plant Varieties and Farmers' Rights Bill, 1999.

[16]   Schrijver, Nico. (1997). Sovereignty over Natural Resources Balancing Rights and Duties. Cambridge, UK: Cambridge University Press.

[17]   TRIPS Agreement: Agreement on Trade-Related Aspects of Intellectual Property Rights, 15 April 1994. *Marrakech. International Legal Materials* 33: 1197

[18]   United Nations General Assembly Resolution 1803 (XVII). 14 December 1962. Permanent Sovereignty over Natural Resources. 1963 *International Legal Materials* 2: 223.xx

[19]   UPOV Convention. (1961). International Convention for the Protection of New Varieties of Plants, 2 December 1961, Paris. UN Treaty Series 815: 89.

[20] UPOV Convention (1991). International Convention for the Protection of New Varieties of Plants. 2 December 1961. As revised on 10 Nov. 1972, 23 Oct. 1978 and 19 Mar. 1991. 1996. Geneva. UPOV Doc. 221(E).

[21] US Patent No. 5,401,504. 28 March 1995. Use of Turmeric in Wound Healing. Philippe Cullet and Jawahar Raja.

[22] US Patent No. 5,827,521. 27 Oct. 1998. Shelf Stable Insect Repellent, Insect Growth Regulator and Insecticidal Formulations Prepared from Technical Azadirachtin Isolated from the Kernel Extract of Azadirachta Indica.

[23] US Patent No. 5,695,763. 9 Dec. 1997. Method for the Production of Storage Stable Azadirachtin from Seed Kernels of the Neem Tree.

[24] World Trade Organization (1997). India—Patent Protection for Pharmaceutical and Agricultural Chemical Products (US complaint), Report of the Panel, 5 Sept. 1997, WTO Doc. WT/DS50/R and India—Patent Protection for Pharmaceutical and Agricultural Chemical Products (US complaint), Report of the Appellate Body, 19 Dec. 1997, WTO Doc. WT/DS50/AB/R.

[25] World Trade Organization (1998). India—Patent Protection for Pharmaceutical and Agricultural Chemical Products (EC complaint), Report of the Panel, 24 Aug. 1998, WTO Doc.WT/DS79/R.

[26] World Trade Organization (2000). The Relationship between the Provisions of the Multilateral Trading System and Multilateral Environmental Agreements (MEAs). WTO Doc. WT/CTE/W/139.

In: Recent Trends in Biotechnology and Microbiology   ISBN: 978-1-60876-666-6
Editors: Rajarshi Kumar Gaur et al., pp. 209-212   2010 Nova Science Publishers, Inc.

ABOUT AUTHORS

About the book: Biotechnology is one of the major technologies of the twenty-first century. Its wide-ranging, multi-disciplinary activities include recombinant DNA techniques, cloning and the application of microbiology to the production of goods from bread to antibiotics. In this edited book, we provide a complete overview of biotechnology. The fundamental principles that underpin all biotechnology are explained and a full range of examples are discussed to show how these principles are applied; from starting substrate to final product. With the use of comparative studies, this book also discusses the legal, agribusiness and public policy issues that connect intellectual property protection with advancements in agricultural biotechnology.

Rajarshi Kumar Gaur: Dr. Rajarshi Kumar Gaur is presently working as Head and Assistant Professor, Department of Science, Mody Institute of Technology and Science (Deemed University), Lakshmangarh, Sikar, Rajasthan. He did his Ph.D on molecular characterization of sugarcane viruses of India. He partially characterized three sugarcane virus viz., sugarcane mosaic virus, sugarcane streak mosaic virus and sugarcane yellow luteovirus. He received MASHAV fellowship in 2004 of Israel government for his post doctoral studies and joined The Volcani Centre, Israel and then shifted to Ben Gurion University, Negev, Israel. In 2007 he received the Visiting Scientist Fellowship from Swedish Institute Fellowship, Sweden for one year to work in the The Umeå University, Umeå, Sweden. He is also a recipient of ICGEB, Italy Post Doctoral fellowship in 2008. He worked on development of marker-free transgenic plant against cucumber viruses. He has made significant contributions on sugarcane viruses and published 25 national/international papers and presented near about 30 papers in the national and international conferences. He has also visited Thailand, New Zealand, London and Italy for the sake of attending the conference/workshop. Recently, he received two projects from Department of Biotechnology, Government of India and Department of Science and Technology, India, New Delhi. He is also a Course Coordinator of short term training program funded by Department of Biotechnology, Government of India. He is also member and reviewer of several national and international scientific societies. Presently, he is working on the characterization of watermelon viruses and RNAi technology.

Pradeep Sharma: Dr Pradeep Sharma is presently working as Senior Scientist (Biotechnology) at Directorate of Wheat Research, Karnal, Haryana, India. He did his Ph.D. in Virology in 2002 from Haryana Agricultural University and worked on cotton leaf curl begomoviruses. He specializes in molecular mechanisms underlying trafficking of viruses, protein-protein interactions, evolution & genetic diversity, RNA silencing based applications and genomics of geminiviruses. Dr. Sharma has made significant contribution in the area of functional characterization of geminiviruses from Aisa and published 38 international and national research papers, several extension papers, invited chapters and reviews. He also has the distinction of receiving numerous honours, international fellowships and national awards in recognition to his excellent academic and research contributions. These include: Young Scientist's Award (Plant Protection Sciences (biannual 2005-2006) of the National Academy of Agricultural Sciences, honoured appreciation certificates by the Japan Society for the Promotion of Science for contribution in Science dialogue mission (2007 & 2008), DST Young Scientist Fellowships (2004, 2005 & 2008), and Pran Vohra award (2008-2009) of the Indian science Congress Association etc. As a recipient of JSPS fellowship, he worked in the Tohoku University, Japan during 2006-2008 and received Postdoc fellowship from Ministry of Agriculture and foreign Affairs, Israel and joined the Volcani Centre, ARO, Israel in 2005-2006. He has worked and visited many pioneer laboratories of UK, Japan, France, the Netherlands, Indonesia, Turkey, and Israel. He is member and reviewer of several national and international scientific societies.

Raghvendra Pratap: Raghvendra Pratap Narayan grew up in Faizabad district of Uttar Pradesh. He received his B Sc and M Sc degree from Allahabad University, Allahabad in year 1999 and 2001 respectively. Later on he joined research in Mycology and Microbiology Laboratory, University of Allahabad, Allahabad. He was awarded by Junior and senior research fellowship of CSIR. He worked on microbial ecology and published several national and international papers in journals of repute and contributed invited chapters to books published from India. Dr Narayan has participated in several national and international conferences. Presently, he is working as a lecturer in microbiology at Department of Biotechnology and Microbiology, Mody Institute of Technology and Science, Deemed University, Sikar Rajasthan

Kanti Prakash Sharma: Kanti Prakash Sharma is currently a lecturer in Biotechnology at Mody Institute of Technology and Science. He focussed on tannin degradation for his doctoral degree and currently involved in identifying the applications of tannins in human health. He received fellowship for his MSc programme from DBT, India and qualified NET examination conducted by CSIR-UGC. He was selected for Biotech Industrial Training programme of BCIL and DBT for six months in the area of molecular diagnosis. He was also selected in refresher course entitled "Advances in Biophysics" organized jointly by IASc, NASI and INSA at CCMB Hyderabad. The author also obtained Summer Research fellowship of IASc, NASI and INSA for two months at IHBT Palampur. He also attended various research conferences and seminars and published papers in reputed journals. Recently he is involved in establishing structural based relationship between tannase and its substrate molecules.

Manshi Sharma: Dr. Manshi Sharma is presently working as Lecturer in Biotechnology at Mody Institute of Technology and Science, Rajasthan, India. She did her Ph.D from University of Bikaner in 2007 on Experimental studies on development of median third eye and regeneration of lens and cornea in mammals. He received a senior research fellowship and worked on Weather based Animal Disease Forecast" (WB_ADF) and "Animal Health Information System through disease monitoring and surveillance. She received several international and national awards during her research work. She published near about 11 papers in national and international repute journals. She is also a member of several societies.

Rajiv Dwivedi: Dr. Rajiv Dwivedi is at present on the Faculty of Arts, Science and Commerce, MITS, Lakshmangarh, Sikar, Rajasthan. He was graduated at Ewing Christian College, Allahabad and later on joined the University of Allahabad for his Masters and D.Phil in Science. He was conferred upon the degree of Doctor of Philosophy in 2007 on the topic 'Physiological and Biochemical studies on *Vigna* species under ultraviolet-B stress'. He received junior and senior research fellowship of CSIR. He has contributed 8 research papers in various national and international journals showing variability in response to various stresses. His work mainly focuses on the antioxidant responses of different plant groups. He has participated in numerous national and international seminars in his field and attended some workshops as well.

INDEX

H

J

K

L

N

T